Atlas of Design

Vol. 2 »

editors
Daniel P. Huffman
Samuel V. Matthews

assistant editor
Martin E. Elmer

published by
nacis
the North American
Cartographic Information Society

Atlas of Design, Volume Two

Third Printing, July 2023

Editors: Daniel P. Huffman and Samuel V. Matthews

Assistant Editor: Martin E. Elmer

Published by the North American Cartographic Information Society, *nacis.org*

The type in this book has been set in Avenir, by Adrian Frutiger.

Printed in the United States of America by John Roberts, Inc., Minneapolis, MN.

ISBN 979-8-9866856-2-5

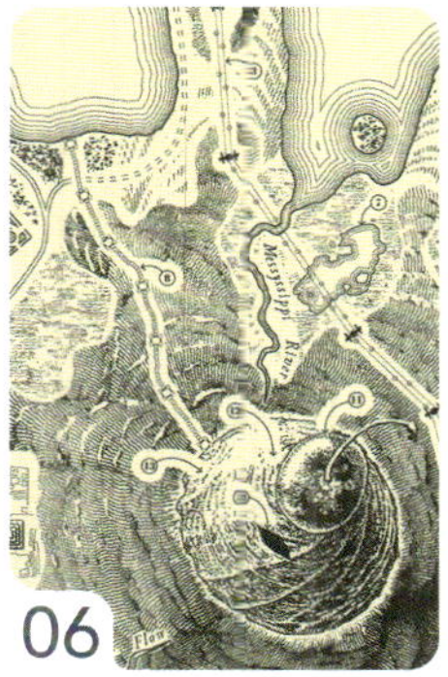
06
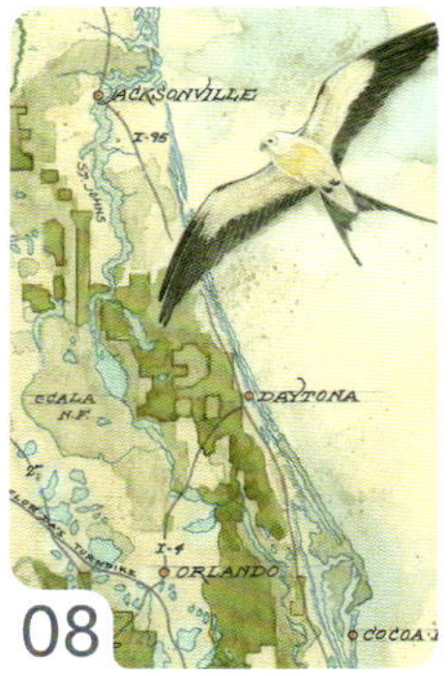
08
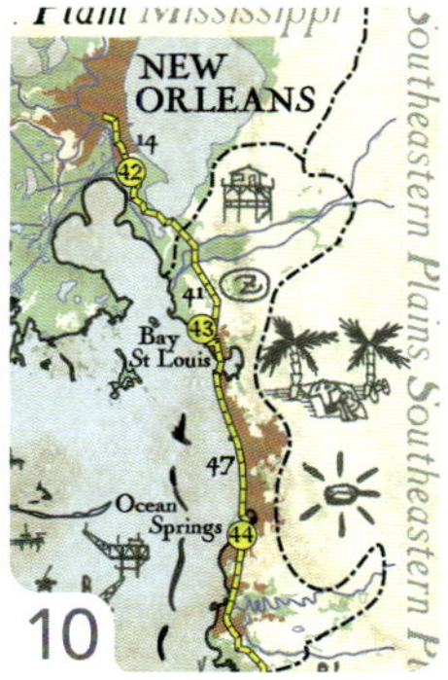
10
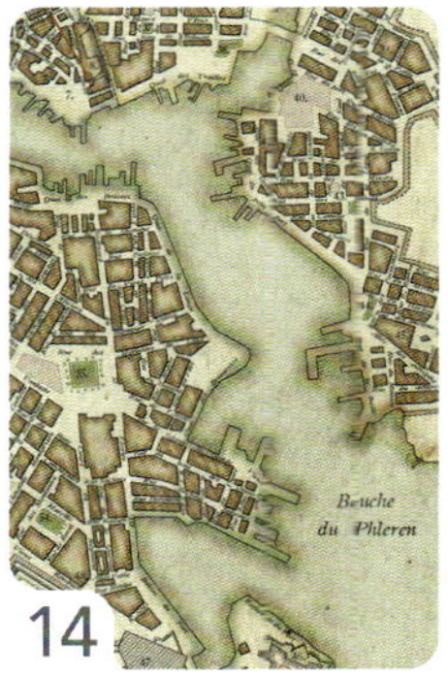
14
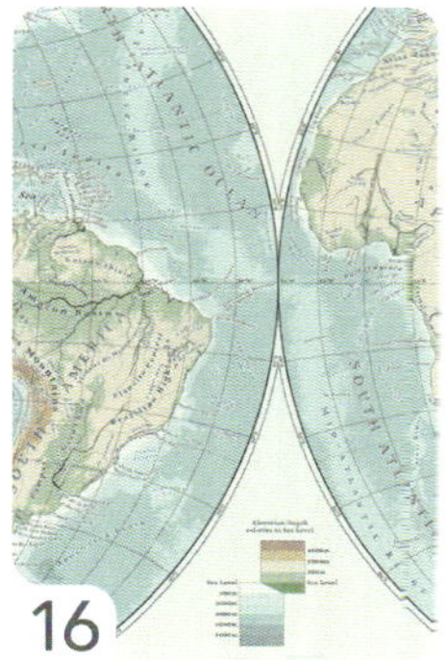
16
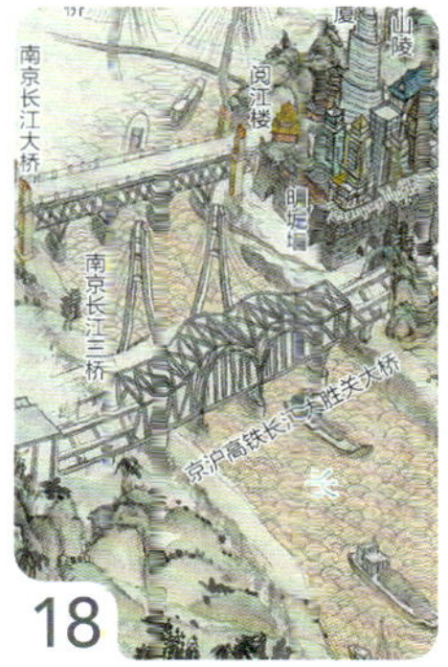
18
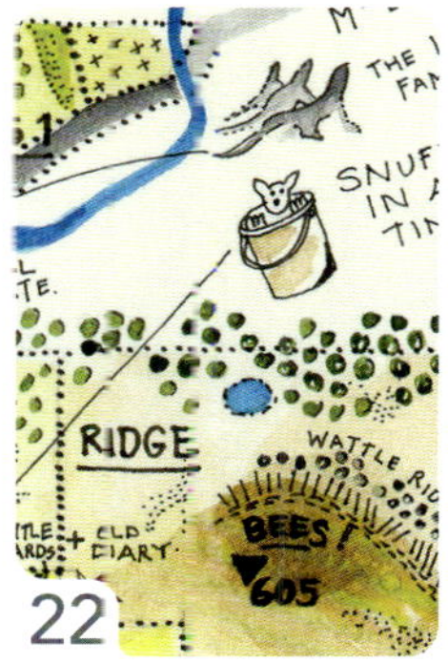
22
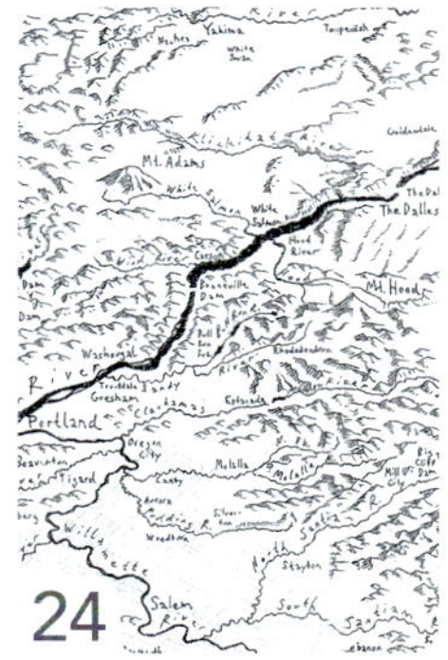
24
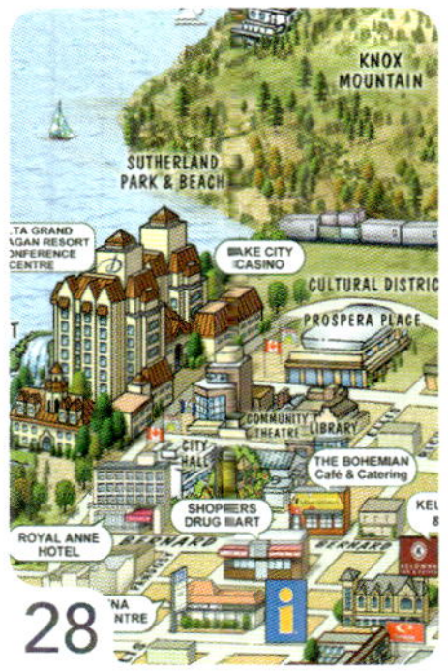
28
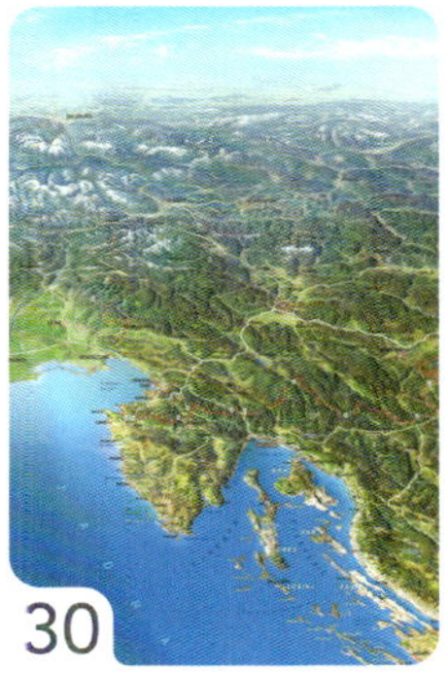
30
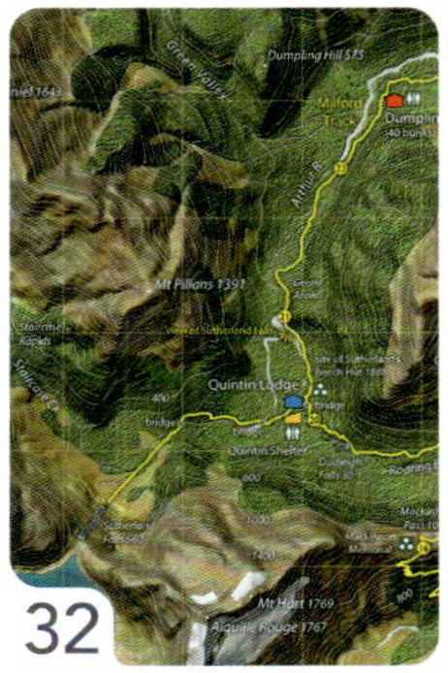
32
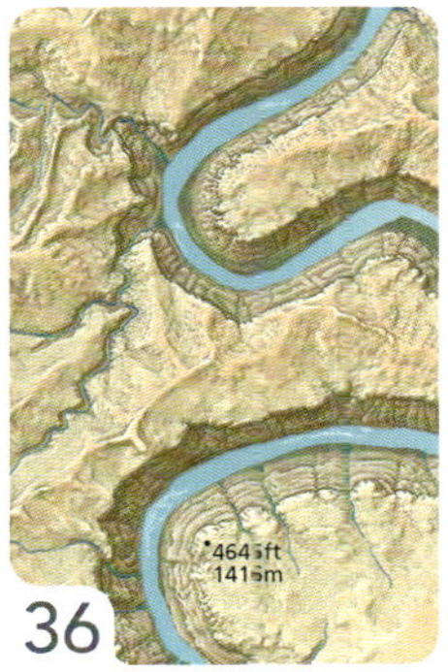
36
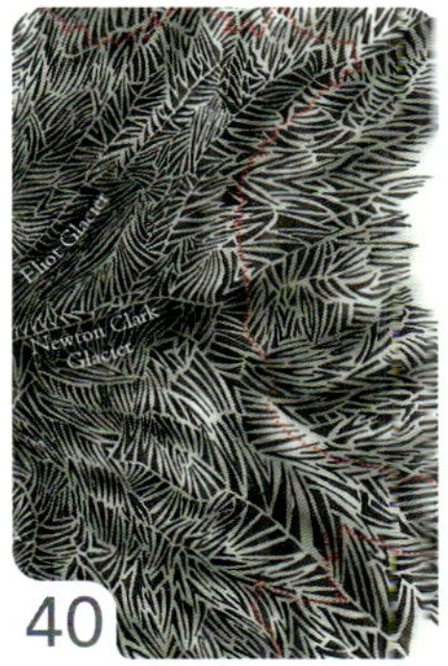
40
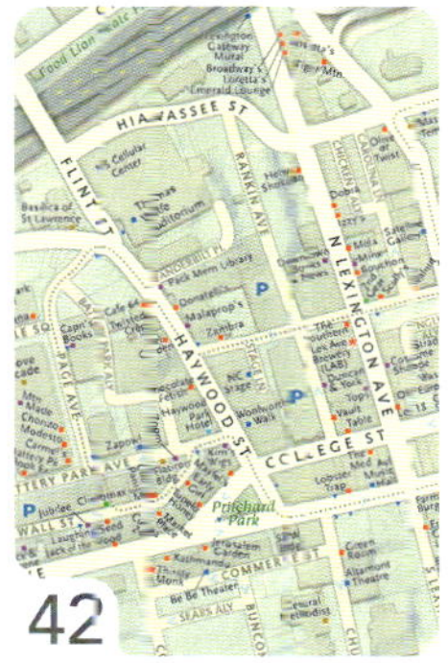
42
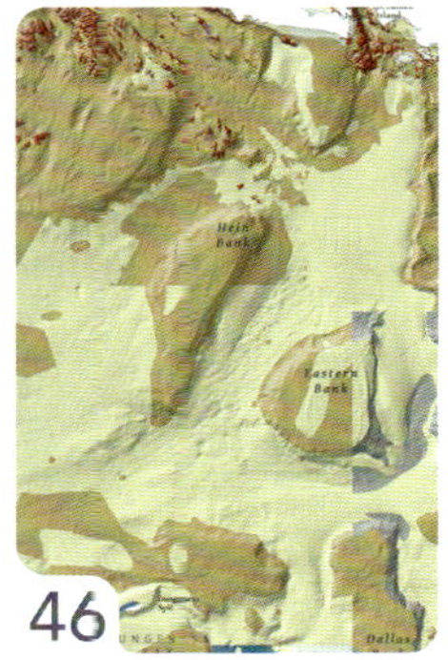
46
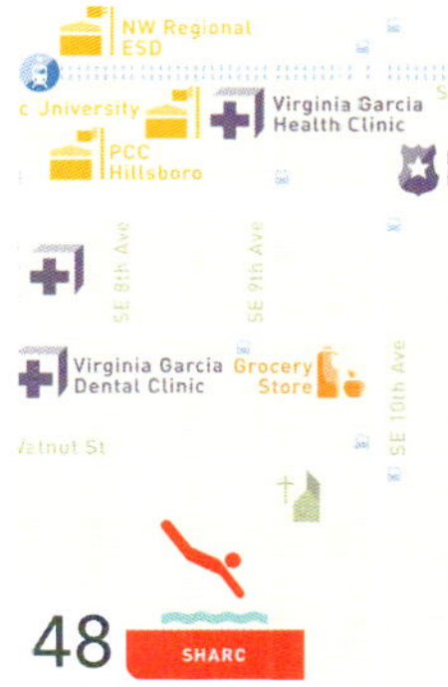
48
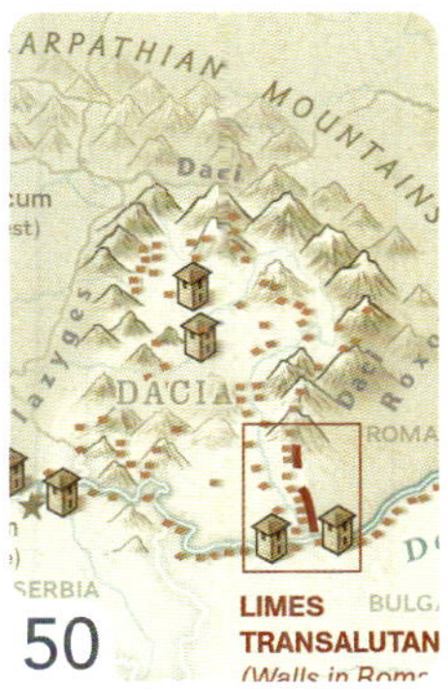
50
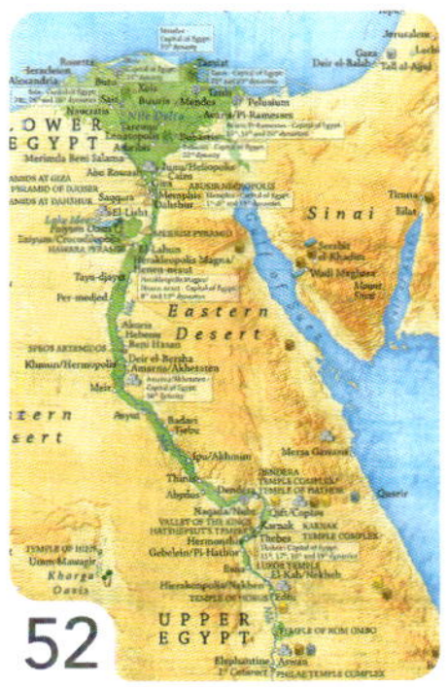
52
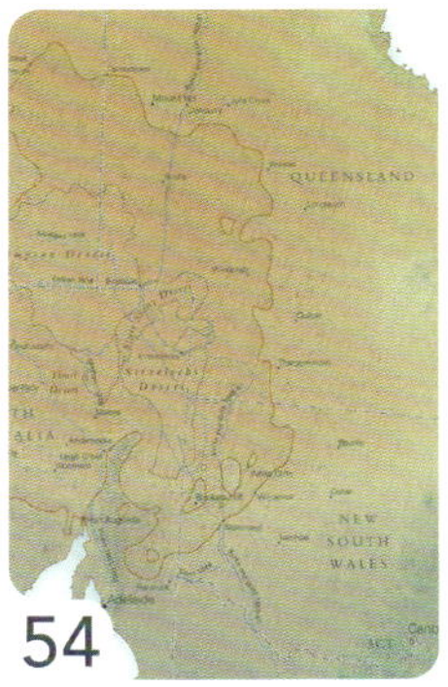
54
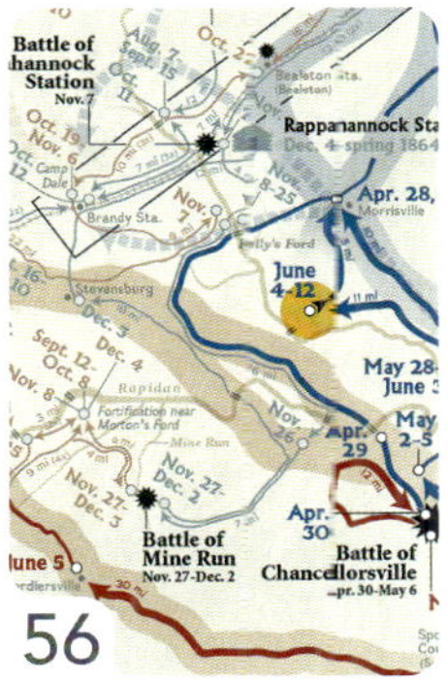
56
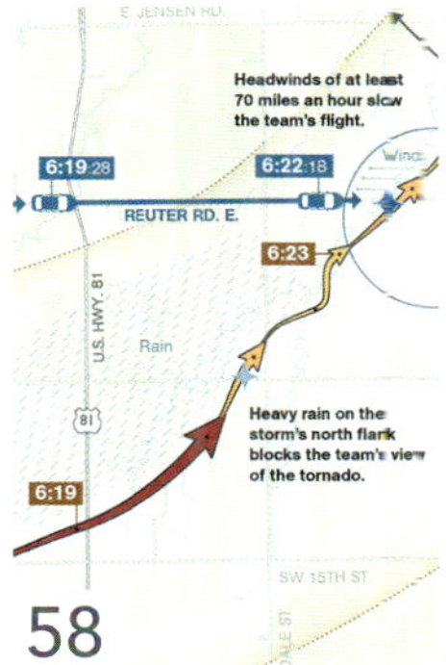
58
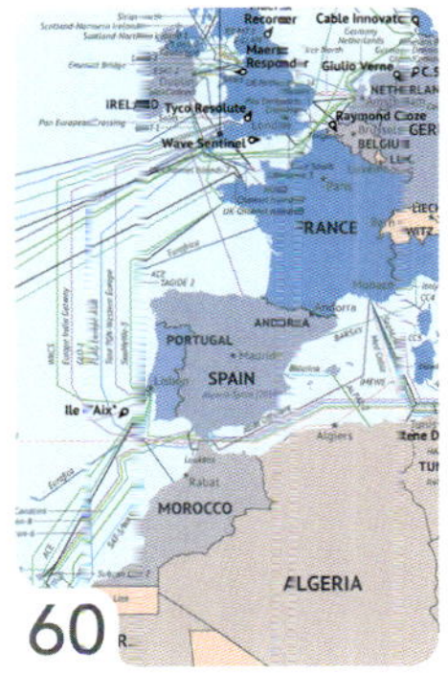
60
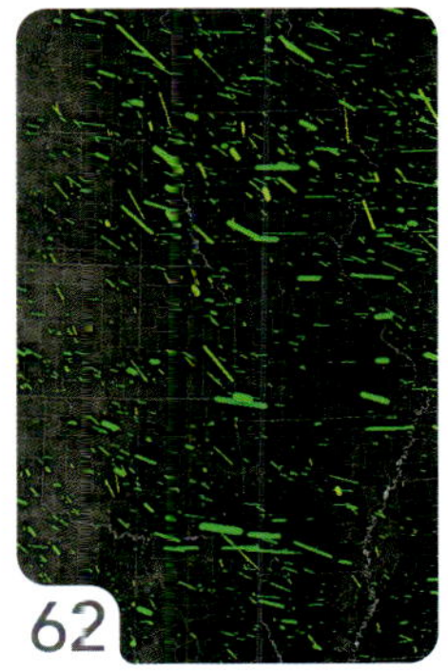
62
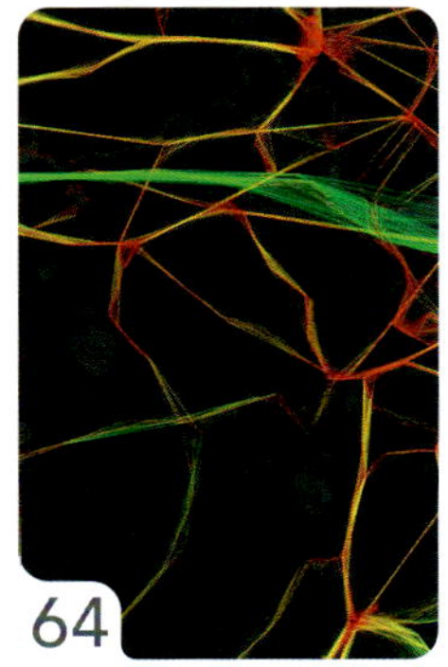
64
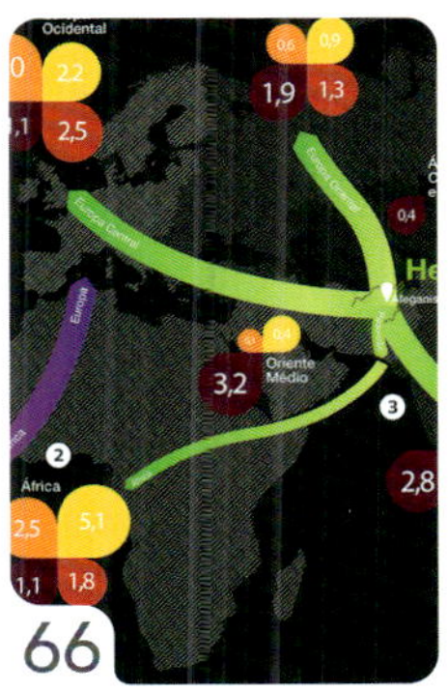
66
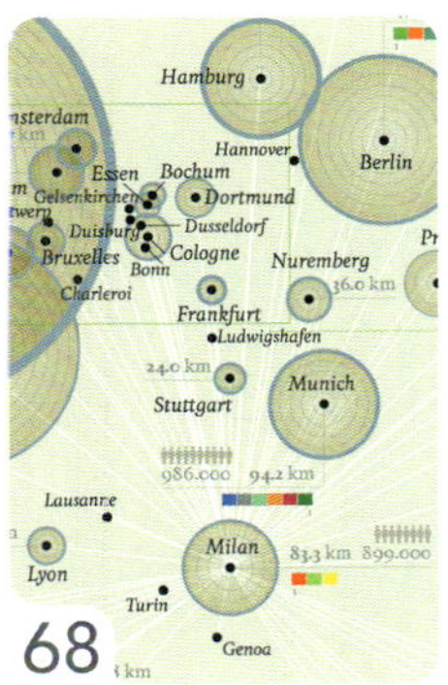
68
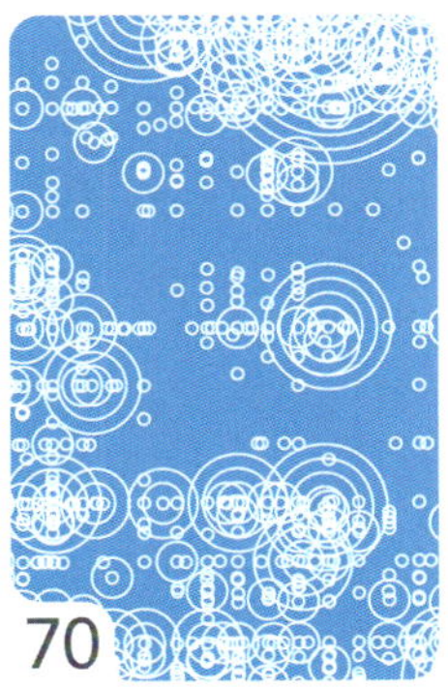
70
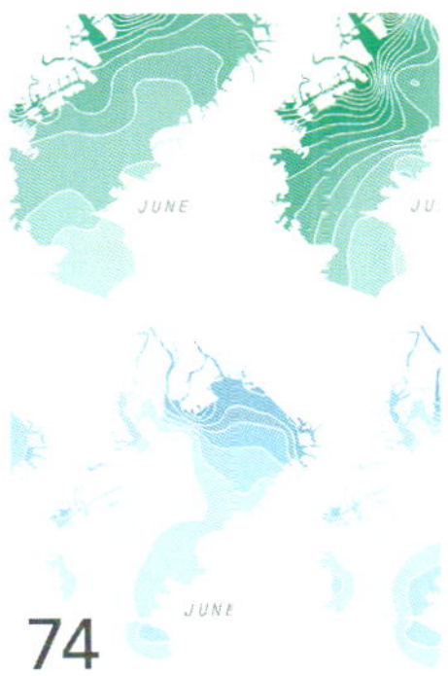
74
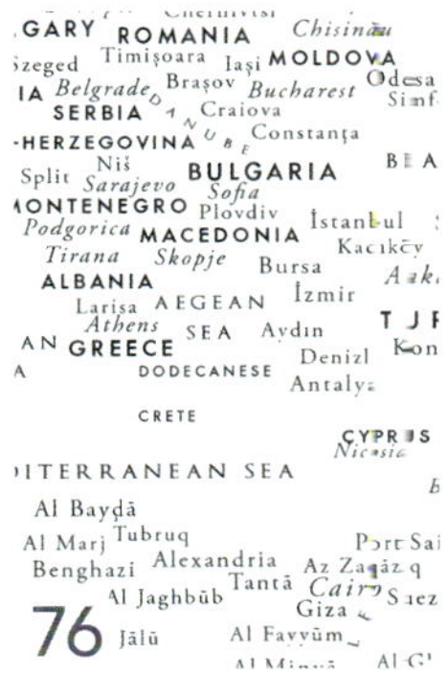
76
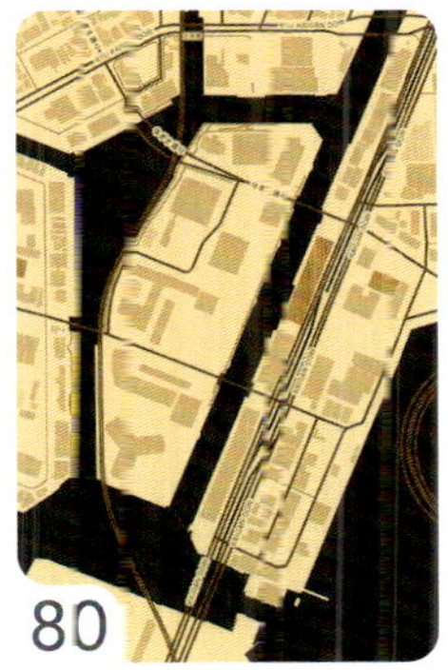
80
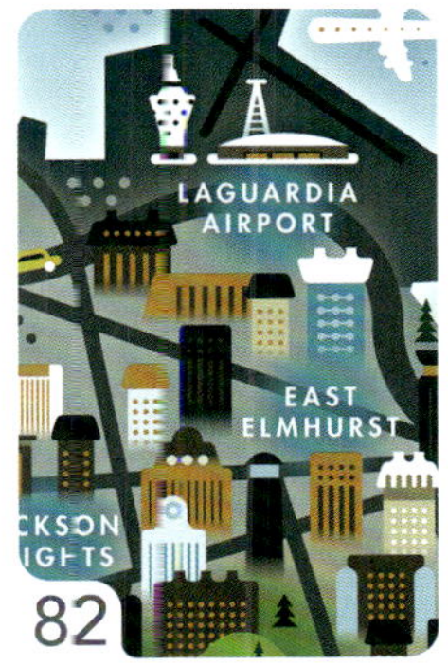
82
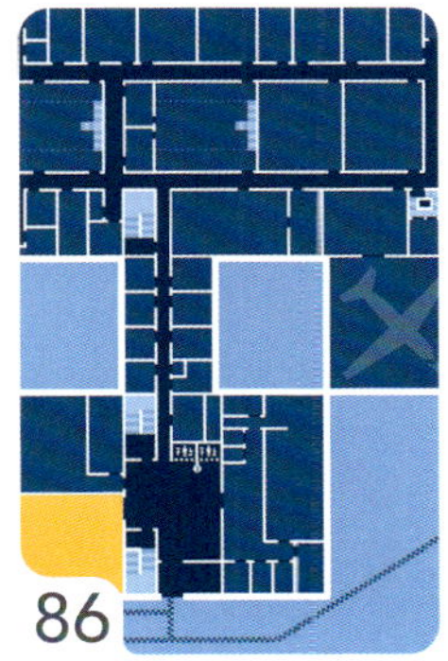
86

Acknowledgements

Though there are only three names on the title page, many hands have been involved in creating this book. Crafting a compilation of some of the best current cartography has been a collaborative effort, and we want to acknowledge our debts to the committed people who helped make it possible.

First, we want to express our gratitude to the NACIS Board of Directors for their continued support for the *Atlas*. Their belief in the publication's importance and value has made this whole endeavor possible.

The first volume of the *Atlas* received over 140 submissions; the size of this response proved that our project was possible, and we are grateful to everyone who took the time to send us their work. This time around, Volume II doubled that total, taking in nearly 300 entries from cartographers spanning 23 countries and five continents. This truly global response has been inspiring, and we want to thank everyone who submitted their work or who spread the word. Even if they don't all have a map featured in this volume, each of them helped make it possible by being willing to contribute to our project.

Such an international array of cartographers brings a wide variety of languages which we, the editors, cannot speak ourselves. We want to give our most humble thanks to the incredibly helpful group of volunteer translators who made it possible to communicate smoothly with contributors around the globe: Costanza Asnaghi, Sarah Bennett, Tiberiu Chelcea, Nadiia Gorash, Veera Helle, Laurent Jégou, Johannes Kröger, Hans van der Maarel, Ramon L. Perez David, Tomás Setubal, daan Strebe, and Xu Wang-Angsüsser.

More submissions take more work to review, and we are exceedingly grateful to the panel of judges that took on the enormous task of scoring them. Their sense of design, keen attention to detail, understanding of geographic data, and experience in the profession helped us narrow down the submissions to a number that could be squeezed into these pages. Our thanks go to AJ Ashton, Cameron Booth, Nat Case, Ginny Mason, Ian Muehlenhaus, Tom Patterson, Margaret Pearce, Rob Roth, Alethea Steingisser, Hans van der Maarel, Tim Wallace, and Andy Woodruff. We should point out that a number of them have maps featured in this book, but they recused themselves from judging their own work, and the final decisions were made by the editors.

For us to put together a book that matched our vision, we needed to think carefully about papers, cover stocks, binding, and more—things which we editors are not experts in. The folks at American Printing have gone above and beyond in patiently teaching us about everything it takes to build a good book.

This volume would not have been possible without the help of many people, and their diligent efforts are apparent in the pages that follow.

The Subjectivity of Cartographic Taste

Kenneth Field

Cartographers have always been storytellers. This metaphor works well for thematic maps, but topographic or reference maps also tell stories: of a landscape, of settlement, and of the shape of the natural and human-modified world that surrounds us. But how you tell the story—the intonation, the language, the structure—is so important to engaging your audience and making them listen. Cartographers take data and wrestle it before applying some graphical treatment that provides a narrative. They codify the story in a visual language that they hope speaks to people. Communicating insights into complex data and revealing patterns is what cartography has always been about, and we are taught early on that the key person in the process of making a successful map is the person that reads it. Your map is likely to fall upon a disinterested audience if you fail to consider whom the map is made for. But when beauty is in the eye of the beholder, how do you craft a map that people see as something pleasing—beautiful even?

Very few would question the value of a well-made map, but just knowing that certain maps work is not much help to anyone trying to master mapmaking themselves. It's an inexact science that is hard to quantify...largely because it's also an artistic endeavour. The mapmaking process is made all the more challenging when we consider aesthetics. Maps are not made in a vacuum, and it is our experiences as scientists, artists, cartographers, and as human beings that shape what we bring to our work. It's inevitable, then, that a practiced cartographer will more likely bring a certain elegance, a visual eloquence, to their work that so often eludes others. That's not in any way to deride laypeople who make their own maps, but rather to encourage them to see something of the artistry and beauty that a professional brings to their work, and perhaps aspire to producing work of a higher caliber themselves. Anyone can, by serendipity, make a beautiful map, but experience, knowledge of the art, imagination, and—on some level—a sense of aesthetics is often seen in more successful work. We should celebrate that work, and encourage wider appreciation of and aspiration towards it.

Exploring the subjectivity of aesthetics and beauty in maps, though, is fraught with danger. We lack a consistent and systemized framework to describe it to someone who can then equally apply it. Maybe that's precisely why some maps are elevated to the pantheon of being considered great while others are considered a little mundane or pedestrian, even if functionally they are no less useful. There is a mighty canon of fantastic cartographic work, but finding ways for people to see it is a challenge in

The breadth of adjectives we use to describe maps is actually rather limited, so trying to define precisely why a map is worthy of being called "well designed" is quite difficult.

an era of sound-bite reportage. To appreciate fine maps, one needs to immerse oneself in them, not just to understand what the map is showing but also what the cartographer was thinking. Why did they make certain decisions? Why is something included or excluded, or shown in a particular colour or style? Simply looking at the map cannot reveal this. You need to step into the map and deconstruct it the way you might deconstruct a Shakespearean play if you are interested in the study of the English language. Understanding the language of the map and the voice in which the cartographer was speaking is no easy task.

Polish-American philosopher Alfred Korzybski's theory of general semantics states that human knowledge is limited by our physical being as well as the structure of language. He has a point, one which resonates with this discussion of aesthetics. Our experience of reality is limited, yet we increasingly see the world through more maps, bigger maps of more data, and more maps of bigger data. It becomes harder to understand how a good map is made, particularly when the mapmaker is also paradoxically less likely to be familiar with the space and place they are mapping themselves, and possibly with cartography also. Korzybski is also remembered for his remark that "the map is not the territory," reflecting the view that an abstraction of reality, derived from something, is not the thing itself, and our reaction to it is not in fact a reaction to reality. This is the very essence of the difficulty in understanding what makes a well-designed map, because we first have to understand the relationship between the work and reality, and the relationship of the cartographer with that reality. By extension, we also have to understand our position in that milieu. We all have different experiences, even when we are experiencing the same thing, and that is why seeing maps in design terms is as much about accepting subjectivity as it is believing the alleged objectivity of the cartographer.

The language we use to describe great maps also underlines subjectivity. We verbalize our own experience by applying labels such as beautiful, well-constructed, aesthetically pleasing, work of art, balanced, etc. The breadth of adjectives we use to describe maps is actually rather limited, so trying to define precisely why a map is worthy of being called "well designed" is quite difficult. At the very most, the experience we get in viewing a map may be shared by one or more other people in part—but likely not in entirety. What works visually for one person may not strike another in the same way—and even if it does, it may get described differently.

As with most artistic endeavors, the world of cartography is also a slave to fashion. None of the maps produced during the Dutch Golden Age were designed with a dark background and translucent linework, yet technology allows this to be a current trend in online mapping. Would we see large ornate letterforms, once the outcome of the copperplate engraving process, on a modern map? Unlikely. Cartography has, at different points, promoted scientific objectivity over artistic representation and vice versa, and how that manifests in cartographic work at different periods in time has also been a function of the technology of map production. Some prefer a rational, scientific approach when translating their experience into map form, with logical, clean lines and a minimum of extraneous detail or artistic flourishes. Others prefer to imbue their work with emotion, and to evoke emotion, and to do so by working outside traditional models of representation. This is an almost philosophical rivalry, yet surely there is a middle ground—one that expects scientific rigor of the underlying data and its manipulation and which is revealed using a strong aesthetic that is seen as appealing and attractive. With a new wave of mapmakers we see new ideas and new fashions, but fashion can be visually disruptive and contentious. This is akin to punk rock challenging light jazz. New mapmakers will likely want to make their mark and be seen as avant-garde, different and free from the shackles of conventional design. New fashions challenge the establishment by making us see things in different ways that counter and modify our own cartographic subjectivity. Making unconventional-locking maps invigorates cartography, and, while we get a lot of woolen neckties, we also get the occasional miniskirt. The former is seen as a fashion faux pas; the latter a classic, timeless design.

Some fashions just don't sit well on a map despite their short-term popularity, but others can begin to define a new genre. Fashion is rarely to everyone's taste, but without diversity and different tastes we revert to bland, one-dimensional dressing of the map. And so fashion goes in cycles. Diversity in the aesthetics of maps is fuelled by more mapmakers coming from fields other than those that deal principally with spatial subjects. Coders, computer scientists, graphic designers, and journalists are all benefiting from tools that have dramatically reduced the barrier to making maps, and they are seeing the value in presenting their information in map form. They often have limited experience of making maps and are unlikely to have formal cartographic training, but this simply brings a different angle to their work. Picasso was formally trained in fine drawing and understood his craft, but he also brought his own perspective to light through his art, which is not to everyone's taste. Is street artist Banksy formally trained? Most biographical pieces suggest not, but his art has captured imaginations and appeals to many people. His style uses a strong and identifiable aesthetic, without which much of his work

Experience, subjectivity, and fashion have a large role to play in the appreciation of many strands of society, and this extends to cartographic work.

would be cast as simply another piece of graffiti. Perhaps there's an overlap between those who enjoy Picasso and those who enjoy Banksy, but that simply demonstrates that variety and alternate visual styles can be seen for their own merit. Maps have style too—and cartographers often develop their own individual style. Raisz, Berann, and Visscher all authored very different maps in unique styles. And what of non-cartographers who have left their own style on the map-making landscape? Think Beck, Snow, Minard.

People's different experiences will lead them to explore and experiment with styles that they themselves are particularly fond of. When we see a body of work, with each entry in a similar vein, we begin to see a trend and a fashion. Some fashions are fads and disappear rapidly; some become slow burners that we all end up appreciating. Some clash with our taste, and some we are happy to revisit from time to time to see if we can repurpose them, finding new uses for techniques that did not quite work in their original context. Experience, subjectivity, and fashion have a large role to play in the appreciation of many strands of society, and this extends to cartographic work. The old maxim of form and function being fundamental to the look of a good map still stands. The data, the message, and the idea in the map must convey something meaningful and have a form that speaks to the audience. The way the map looks is a very important component of how it is understood, even though that look may change over time as new fashions come to the fore. Everything old eventually becomes new again—just with a fresh coat of paint for new eyes. But it is the paint that the cartographer applies that illustrates their careful thought and craft. Every mark on the map has been subject to exhaustive thought and iteration. Genuinely good maps, pleasing maps, and fashionably great maps have some of the sweat and character of their maker in there somewhere. Their experience in making the map should make our experience of reading it pleasurable.

The map may not be the territory, but as far as cartography is concerned, maps are firmly part of the territory. They give us a sense of the world because they are part of the world and they excite our senses. They also help the continued development of cartography and become part of its ever-changing face. Fashion and change make for a vibrant world of maps, and one to rejoice in. This atlas does just that. I am quite sure every reader will appreciate and connect with many maps in this volume, but isn't it interesting that possibly not all maps will be to everyone's taste? That is why cartography is so fascinating and artistically invigorating.

Ice Cream Island »

Chris ***Brackley*** *(Cartographer), John* ***Blas****, and Derek* ***Shields*** *(Creative Directors)*

> *Like all maps of secret or unknown lands, this map was designed to spark the imagination.*

A child will discover this map on the inside of the brown paper wrapping when they first open their copy of the children's novel *The Legend of Ice Cream Island: The First Cold War.* The book itself looks like an ice cream sandwich, and the map is an enticement to imagine what lies between the chocolate wafer covers. As the child reads the novel, the map serves to orient and enrich their mental picture of the setting in which this fanciful story unfolds.

Like all maps of secret or unknown lands, this map was designed to spark the imagination. And what better to serve this purpose than a cartographic style that brings to mind a time when there really was a "great unknown," when the map of the world wasn't yet filled in?

The layout and design elements of *Ice Cream Island* are borrowed from eighteenth- and nineteenth-century cartographic design. It includes a liberal dose of ornate design elements, from the legend boxes to the compass rose, which serve not only to balance the negative space but to add texture and beauty (not to mention reinforcing the ice cream theme). The whimsical features of the map borrow from real geography (Madagascar's real shape for the island of Splatagascar), as well as pure fantasy (splatapults and bananamarans).

Some of this map was rendered with pen and ink (the central ice cream cone with cherry on top), some with computer-drawn lines (water, shoreline, roads, and places of interest), and some by cutting, pasting, and warping photoshopped pieces of actual antique maps (hachure marks and forest textures). Nothing says "old map" like hachure marks (the parallel linework that denotes hills, or, in this case, melting ice cream). The authenticity of this map is rooted in these hachure marks, and while for now I've had to steal them from the ink-work of long-dead cartographers, someday I hope to learn the lost art of rendering them with my own pen.

1. SPLATAGASCAR
2. WHIPPED CREAM CLOUDS WITH RAINBOW SPRINKLES
3. CONEDOLA
4. SWIZZLE STICK PARK
5. SPLATAPULT FIELDS
6. FROZE GARDEN
7. CHOCOLAKE
8. THE GRAPE WALL OF CHINA
9. SPILLUS ISLAND
10. DROPSICLE FOREST
11. CHERRY ON TOP
12. UPPER SPLATTERHORN
13. LOWER CONEACOPIA
14. BRAINFREEZE'S LAIR

15. BEACHES & CREAM
16. MOCHI PICHU
17. 21 CONE SALUTE
18. SLAMWICH PARK
19. CHOCOLATE BANANA BRIDGE
20. THE LEANING TOWER OF GELATO
21. SCOOPSADAISY
22. BANANAMARANS

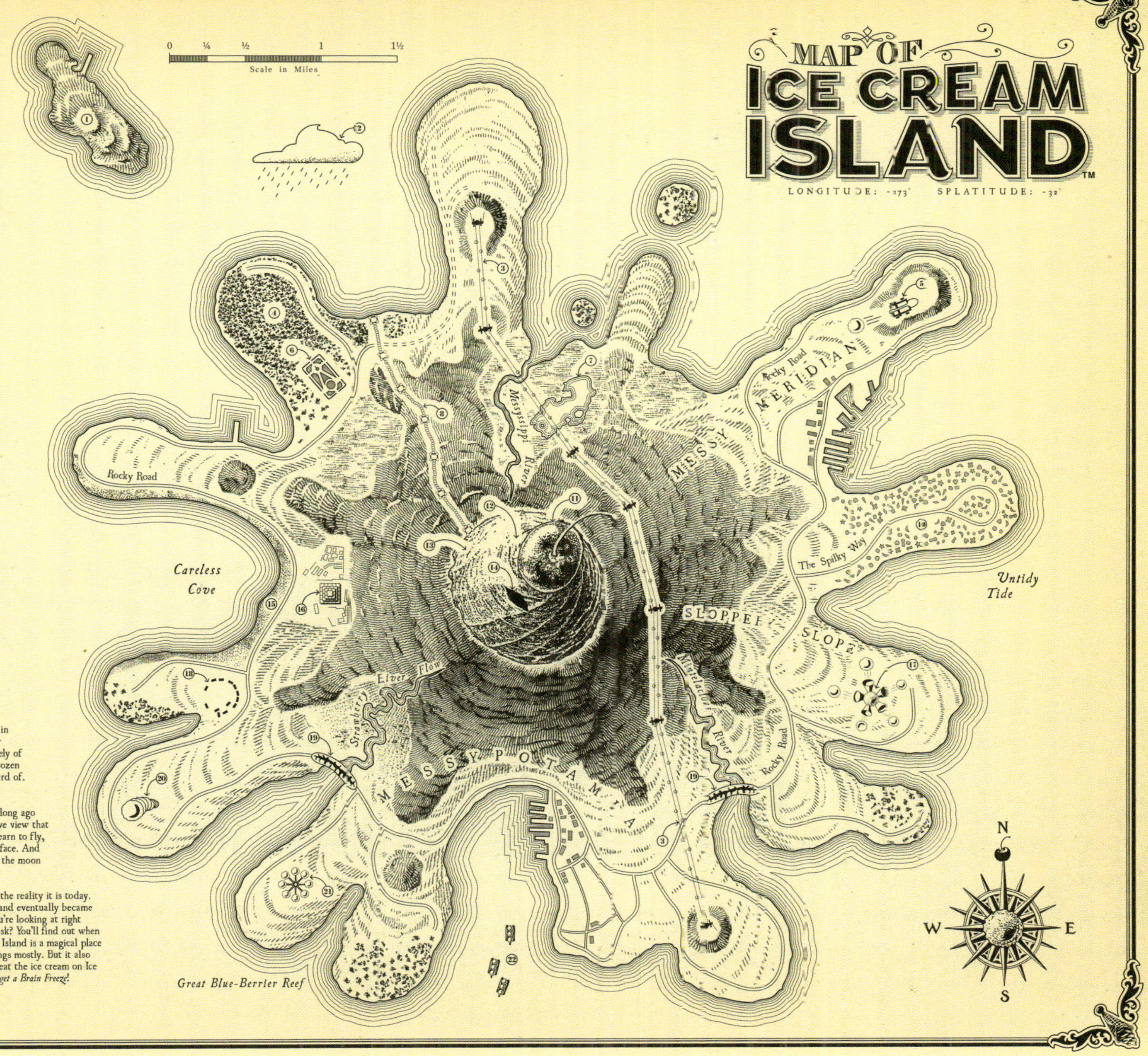

It's been said that Ice Cream Island exists only in your imagination. That it's not really *real*. How could such a place exist? An island; made entirely of ice cream; with a *Population* made entirely of frozen confections; floating in the open ocean? Unheard of. Preposterous. Utter nonsense!

Perhaps. What's *not* nonsense, however, is that long ago many smart people loudly voiced their collective view that man would never fly – when not only did we learn to fly, we flew to the moon and walked its barren surface. And it's absolutely positively certain that journey to the moon began *first* in someone's imagination.

That's precisely how Ice Cream Island became the reality it is today. It started as an idea in someone's imagination, and eventually became the legend that this book and this map that you're looking at right now are based upon. Who's imagination you ask? You'll find out when you read the book. Suffice it to say, Ice Cream Island is a magical place with amazing things to experience – sweet things mostly. But it also has secrets and just one single rule. DO NOT eat the ice cream on Ice Cream Island too fast, *or you will most assuredly get a Brain Freeze!*

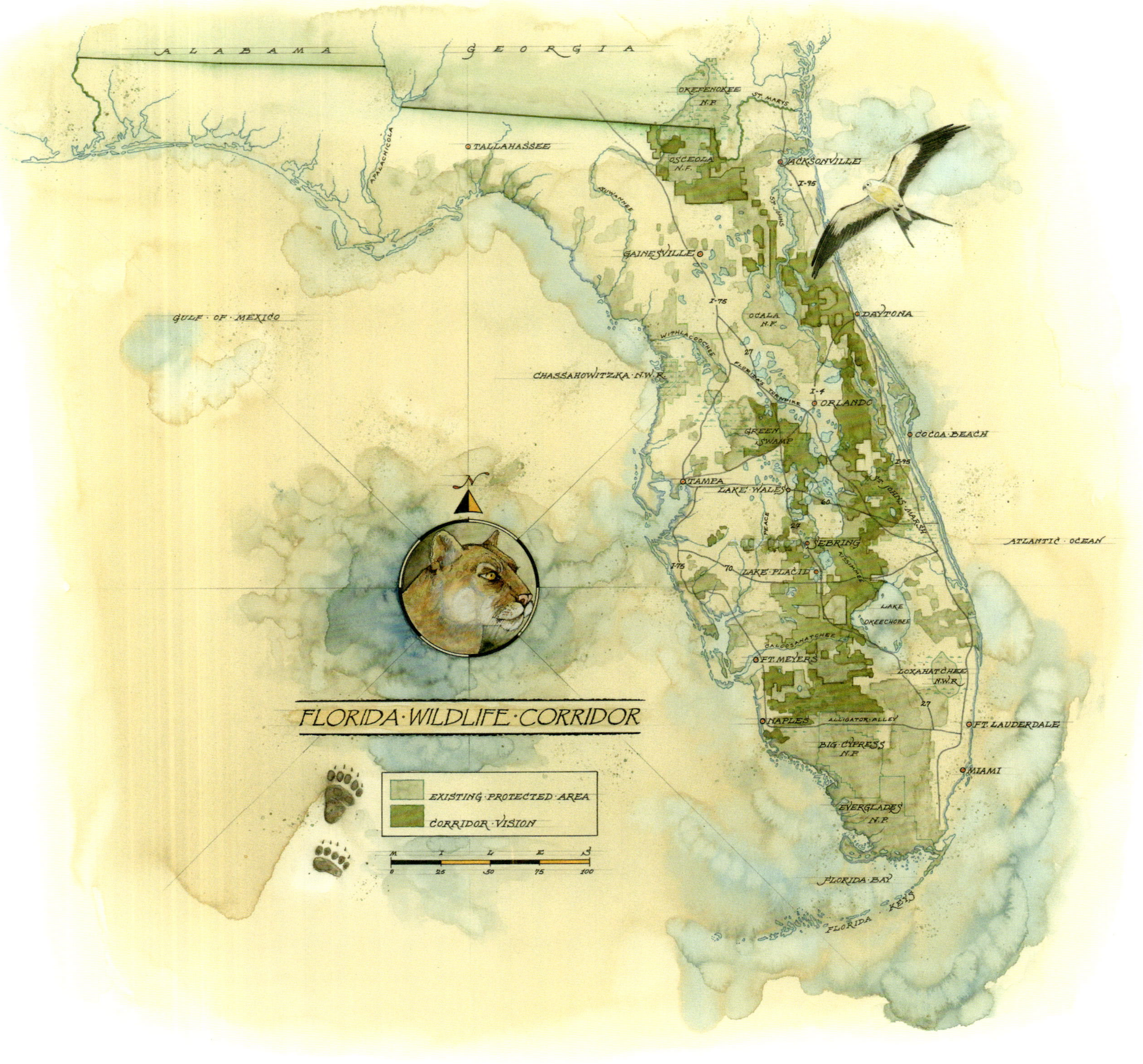
FLORIDA · WILDLIFE · CORRIDOR
ALABAMA
GEORGIA
OKEFENOKEE N.P.
ST. MARYS
TALLAHASSEE
JACKSONVILLE
OSCEOLA N.F.
APALACHICOLA
SUWANNEE
I-95
GAINESVILLE
I-75
OCALA N.F.
DAYTONA
GULF · OF · MEXICO
WITHLACOOCHEE
FLORIDA'S TURNPIKE
CHASSAHOWITZKA · N.W.R.
I-4
ORLANDO
COCOA · BEACH
GREEN SWAMP
TAMPA
LAKE · WALES
ST. JOHNS · MARSH
SEBRING
ATLANTIC · OCEAN
LAKE · PLACID
LAKE OKEECHOBEE
CALOOSAHATCHEE
FT. MEYERS
LOXAHATCHEE N.W.R.
NAPLES
ALLIGATOR · ALLEY
FT. LAUDERDALE
BIG · CYPRESS N.P.
MIAMI
EXISTING · PROTECTED · AREA
CORRIDOR · VISION
EVERGLADES N.P.
M I L E S
0 25 50 75 100
FLORIDA · BAY
FLORIDA · KEYS

« Florida Wildlife Corridor

Mike **Reagan**

My assignments for magazines and books have allowed me to document and sometimes witness not only war-ravaged countries suffering through seemingly unending violence, political instability, and ethnic strife, but also the environmental and economic pressures on communities, languages, and cultural icons around the world. My maps are all hand painted in watercolor and hand lettered in ink. I never use computers because to my mind, a hand-painted map has a certain organic human element that people sense and are naturally drawn to. Throughout my career, as many ancient mapmakers before me have done, I've approached mapmaking as a sort of storytelling, a way of eliciting mood, inspiration, and imagination. As with the early mapmakers, my hand is intentionally evident and my work tries to connect the dots of our global community.

The challenge in painting the *Florida Wildlife Corridor* map was to make it visually interesting enough to draw people in for a closer look and give them a better understanding of the scope of the Corridor project. The purpose was to show the Corridor as the project continues to connect remaining natural lands, waters, working farms, forests, and ranches from the Everglades north to Georgia and west to Alabama, protecting a functional ecological corridor for the health of people, wildlife, and watersheds. The Corridor addresses the fragmentation of natural landscapes and watersheds from the Everglades ecosystem north. Contributing to the fragmentation problem is the disconnect between the perceptions of Floridians and the real need to keep natural systems connected. The Florida Wildlife Corridor is positioned to mend the perception gap through an education and awareness campaign that demonstrates the connection between the landscapes and watersheds.

Whether in times of war or peace, no matter where in the world, I've always tried to see and paint the beauty of the oceans and rivers, mountains and islands in these faraway and often exotic places—but in the end, for me at least, the subject is the innate beauty of watercolor paint itself.

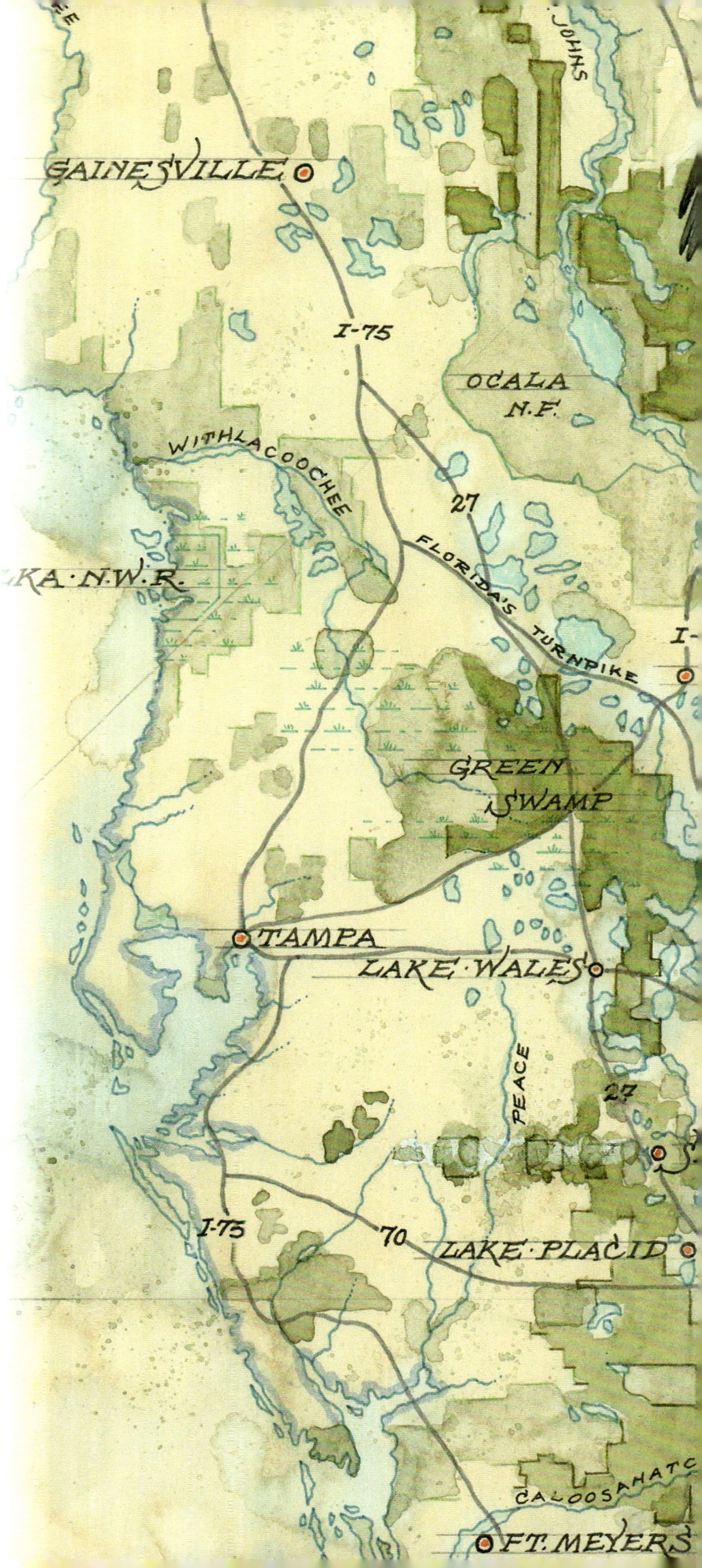

» *...I wanted to create a map with a soul that illustrated the beauty and chaos of the road.*

The Road from Madison »

Andrew **Umentum**

For five months last year I travelled by bicycle through the heart of America, from the Great Lakes down to the Gulf Coast and back, in order to expand my understanding of this country and test my knowledge of how it functions. My two companions and I were directed by circumstance and curiosity, which led to an entirely unique perspective that I attempted to capture in this work.

Balance is essential to all things—biking is an obvious example, but good design relies even more on that principle. I wanted to juxtapose the quantifiable aspects of the trip with the personal experiences that gave life to the places we rode through. The routes and mileage alone are devoid of meaning without the illustrations that trigger the memory of what those lines and numbers represent. The spatial accuracy of the shapefiles is balanced against the hand-drawn and unscaled symbols of place which, like the journey itself, are highly subjective and dependent on the traveler.

The simplicity in color allows for complexity in linework and gives an antique aesthetic most associated with the Age of Exploration. The entire map, including breaking the journey into twelve strips of varying orientations, was directly influenced by *Britannia*, John Ogilby's 1675 road atlas of England & Wales. The notion of framing a place according to artistic principles and utility, versus cartographic orthodoxy, is a defining factor of Ogilby's work as well as this map. The dynamic compass rose is true to the nature of travel and allows for continuity of motion from top to bottom between frames.

Overall I wanted to create a map with a soul that illustrated the beauty and chaos of the road. The typical road atlas has sanitized the landscape and sucked all life out of place, which I believe has a direct impact on the degradation and homogenization of culture. As Americans we constantly define place and ourselves based on arbitrary boundaries of statehood, and continue to ignore the land and people that truly differentiate this country. I hope this map can inspire a more balanced cartography.

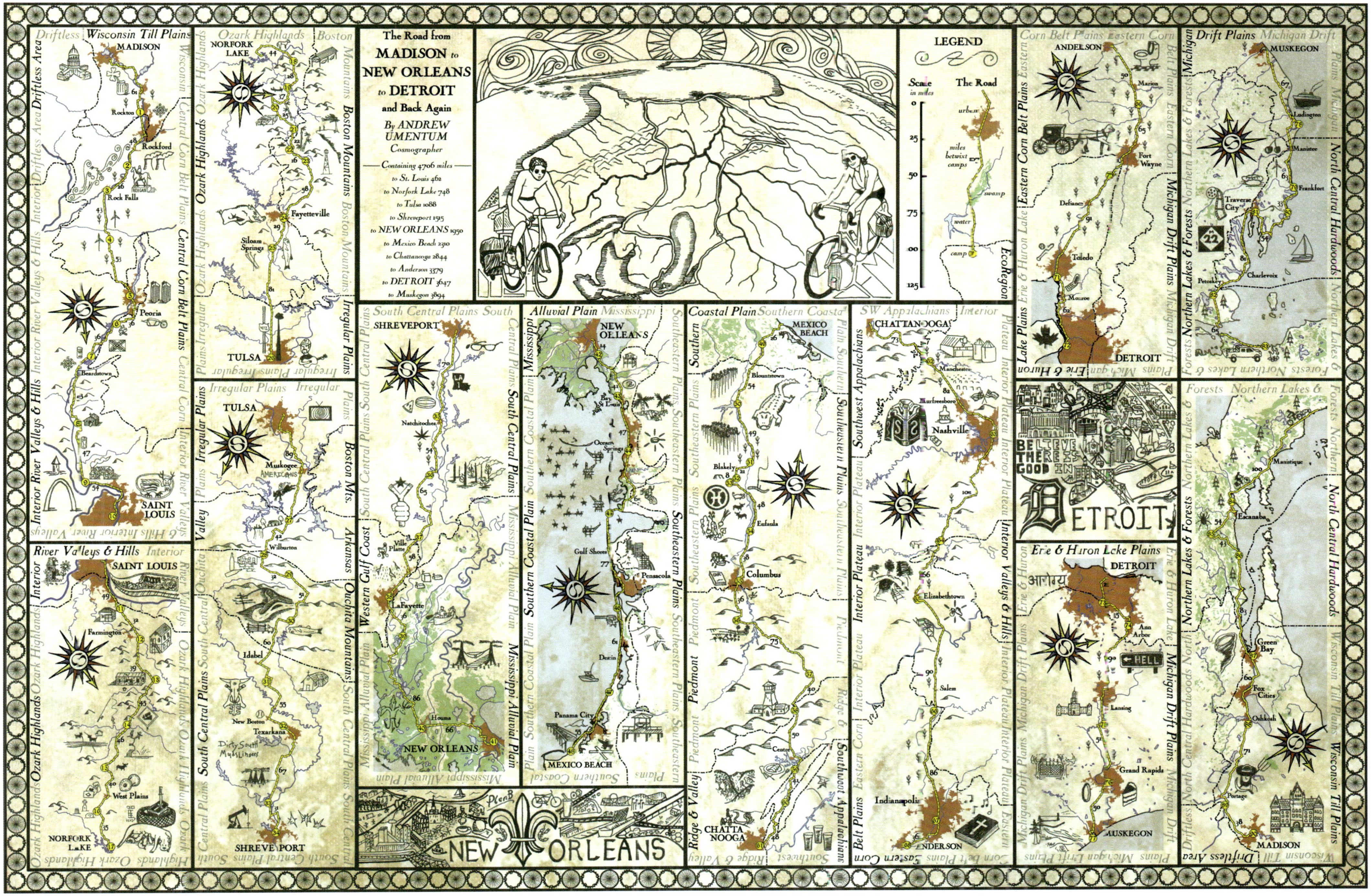

The Road from
MADISON to
NEW ORLEANS
to DETROIT
and Back Again
By ANDREW UMENTUM
Cosmographer
Containing 4706 miles
to St. Louis 462
to Norfork Lake 748
to Tulsa 1088
to Shreveport 1515
to NEW ORLEANS 1950
to Mexico Beach 2310
to Chattanooga 2844
to Anderson 3379
to DETROIT 3647
to Muskegon 3894
LEGEND
Scale in miles
The Road
urban
miles betwixt camps
swamp
water
camp
EcoRegion
Wisconsin Till Plains
MADISON
Rockton
Rockford
Rock Falls
Peoria
Beardstown
SAINT LOUIS
Driftless Area
Interior River Valleys & Hills
Central Corn Belt Plains
Ozark Highlands
NORFORK LAKE
Fayetteville
Siloam Springs
TULSA
Boston Mountains
Irregular Plains
Farmington
West Plains
Muskogee
Wilburton
Idabel
New Boston
Texarkana
SHREVE PORT
Ouachita Mountains
Arkansas
South Central Plains
SHREVEPORT
Natchitoches
Ville Platte
LaFayette
Houma
NEW ORLEANS
Western Gulf Coast
Mississippi Alluvial Plain
Alluvial Plain
Bay St. Louis
Ocean Springs
Gulf Shores
Pensacola
Destin
Panama City
MEXICO BEACH
Southern Coastal Plain
Southeastern Plains
Coastal Plain
Blountstown
Blakely
Eufaula
Columbus
Centre
CHATTA NOOGA
Piedmont
Ridge & Valley
Southwest Appalachians
SW Appalachians
CHATTANOOGA
Manchester
Murfreesboro
Nashville
Elizabethtown
Salem
Indianapolis
ANDERSON
Interior Plateau
Interior Valleys & Hills
Eastern Corn Belt Plains
Corn Belt Plains
ANDERSON
Marion
Fort Wayne
Defiance
Toledo
Monroe
DETROIT
Erie & Huron Lake Plains
Lake Plains
BELIEVE THERE IS GOOD IN DETROIT
DETROIT
Ann Arbor
HELL
Lansing
Grand Rapids
MUSKEGON
Michigan Drift Plains
Drift Plains
MUSKEGON
Ludington
Manistee
Frankfort
Traverse City
Charlevoix
Petoskey
North Central Hardwoods
Northern Lakes & Forests
Manistique
Escanaba
Green Bay
Fox Cities
Oshkosh
Portage
MADISON
Wisconsin Till Plains
NEW ORLEANS

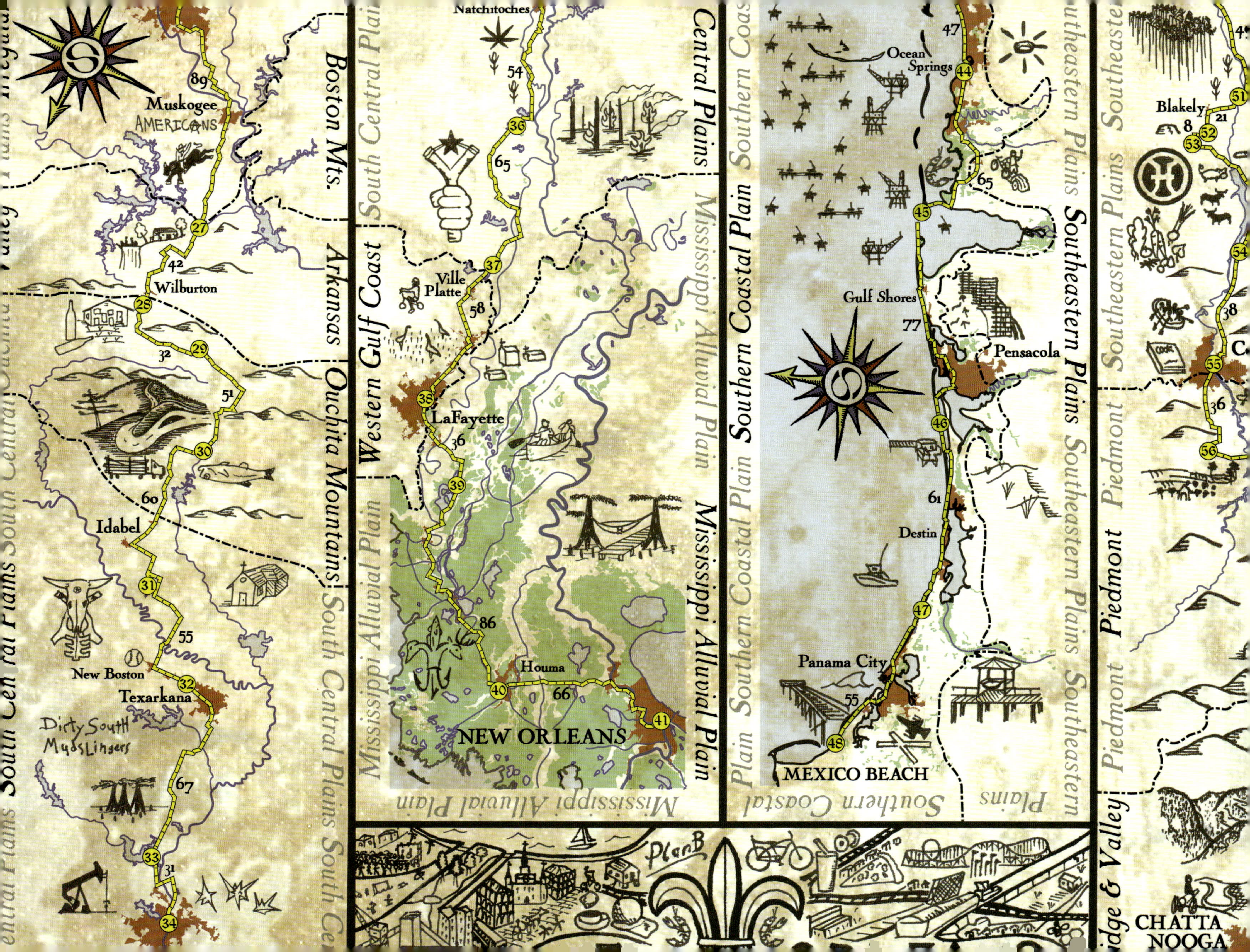
Boston Mts.
Arkansas
Ouchita Mountains
South Central Plains
Muskogee
AMERICANS
Wilburton
Idabel
New Boston
Texarkana
Dirty South
MudSlingers
Natchitoches
Western Gulf Coast
Ville Platte
LaFayette
Houma
NEW ORLEANS
Mississippi Alluvial Plain
Central Plains
Southern Coastal Plain
Ocean Springs
Gulf Shores
Pensacola
Destin
Panama City
MEXICO BEACH
Southeastern Plains
Blakely
Piedmont
CHATTA
NOOGA
PlanB

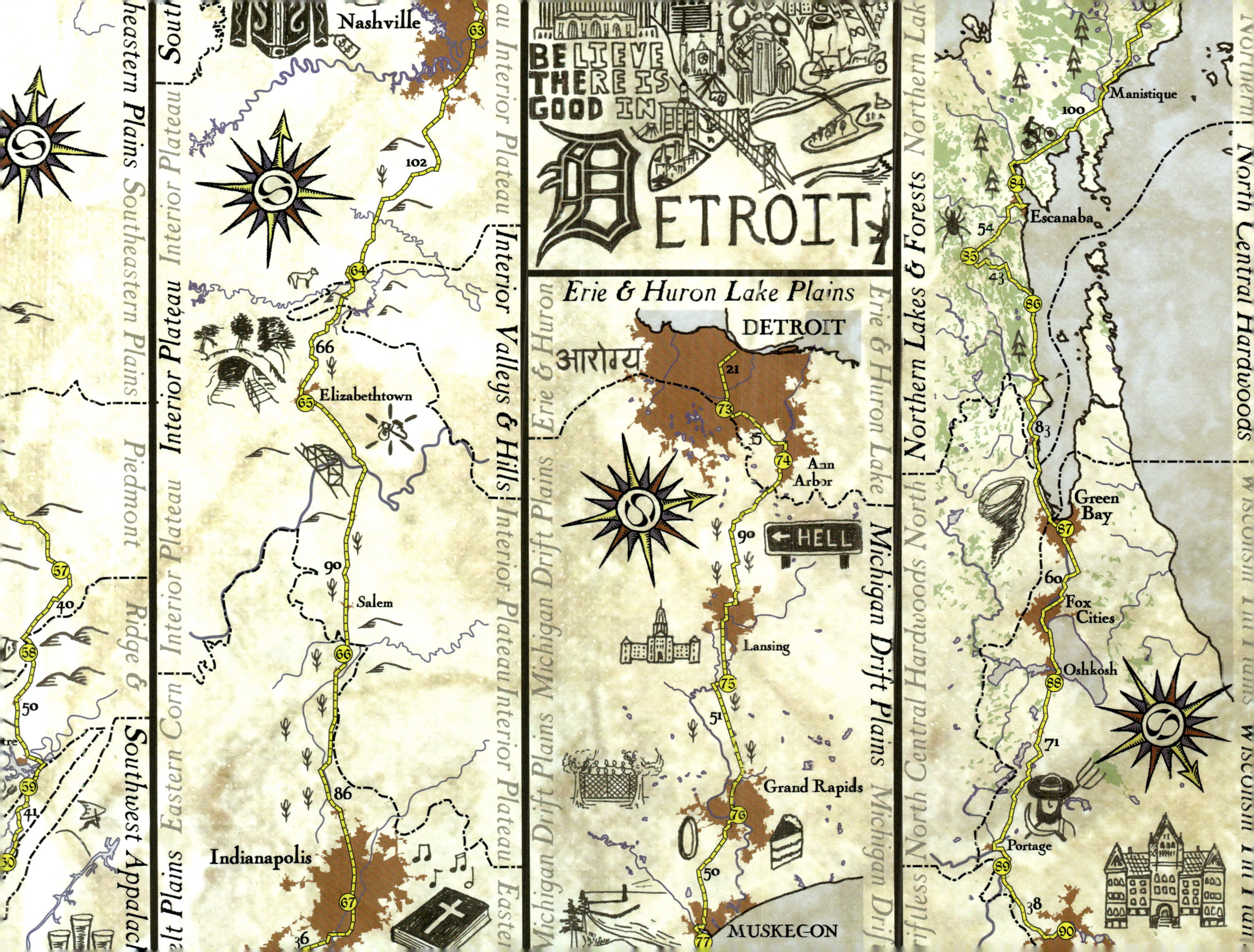
Southeastern Plains
Piedmont
Ridge &
Southwest Appalac
57
40
58
50
59
41
Interior Plateau
Eastern Corn
Nashville
63
102
64
66
Elizabethtown
65
90
Salem
66
86
Indianapolis
67
36
Interior Valleys & Hills
Interior Plateau
BELIEVE THERE IS GOOD IN
DETROIT
Erie & Huron Lake Plains
DETROIT
आरोग्य
21
73
35
74
Ann Arbor
90
HELL
Lansing
75
51
Grand Rapids
76
50
MUSKEGON
77
Michigan Drift Plains
Northern Lakes & Forests
North Central Hardwoods
Manistique
100
84
Escanaba
54
35
43
86
83
Green Bay
87
60
Fox Cities
Oshkosh
88
71
Portage
89
38
90

Yphyrion

cartographiée en l'an 157 du nouveau calendrier de l'Empirium

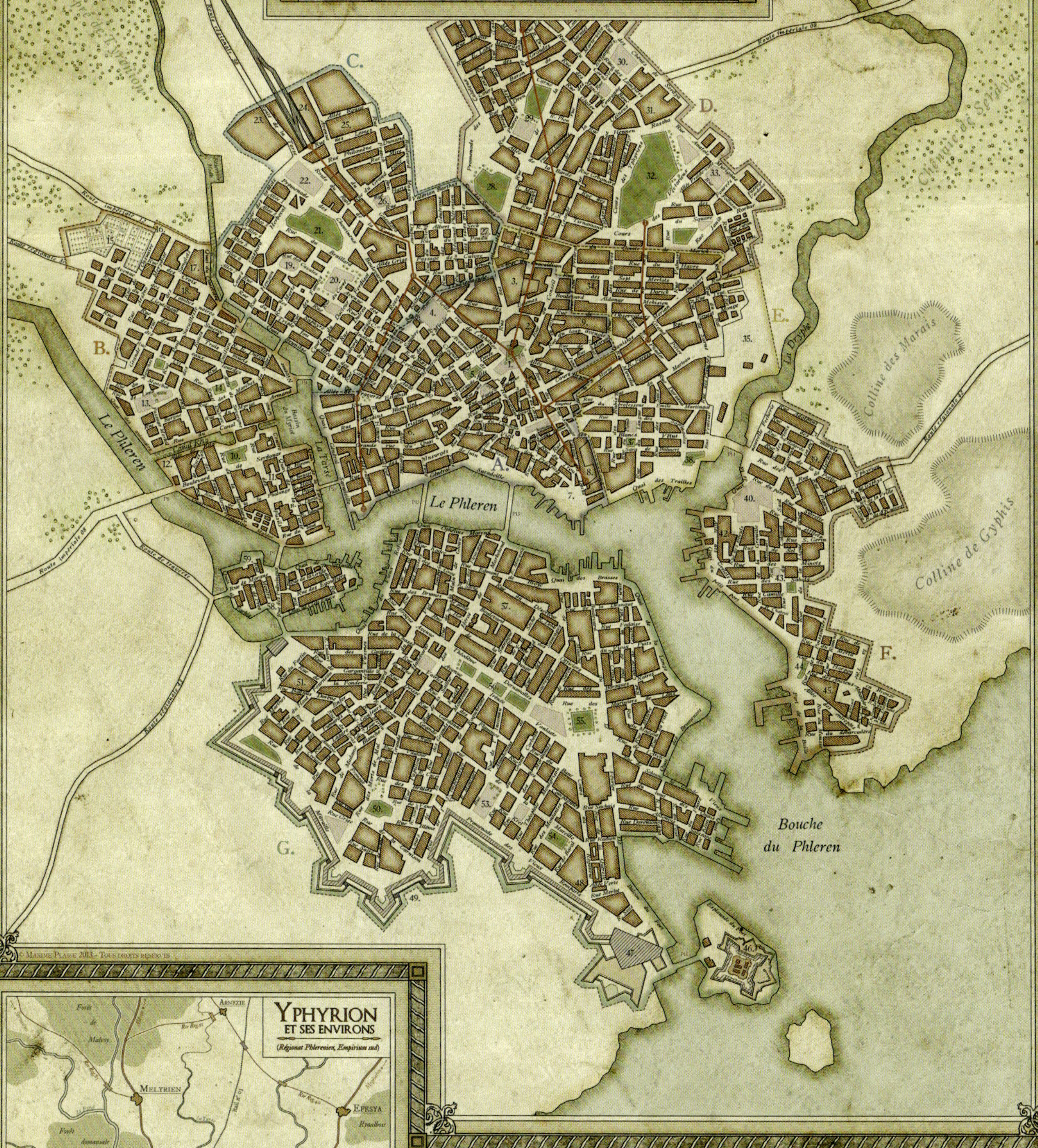

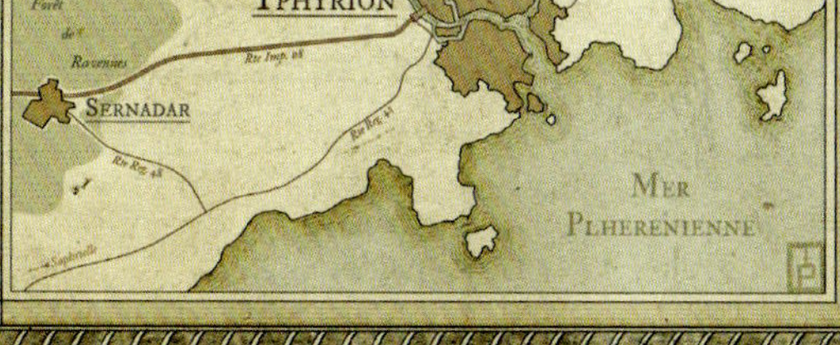

District A.
1. Place de l'Horologium
2. Cercle de l'Horologium
3. Praesidium du Conseil
4. Place d'Outre-monde
5. Parc doré
6. Université de Sabrenet
7. Grands docks du Phleren
8. Pecherie du Trou
9. Ateliers du Loup

District B.
10. Jardins de la Bombarde
11. Marché aux greffons
12. Pointe de Retz
13. Place de la lune
14. Square du centre
15. Cimetière de Modh
16. Entrée du vieil ossuaire
17. Maison funéraire
18. Tribunal Montdroux

District C.
19. Place des hussards de Rhen
20. Place du carré
22. Terrain des chemineux
23. Ateliers du rail
24. Gare Yphyrienne
25. Entrepots rouges
26. Place de Bro
27. La potence

District D.
28. Parc des songes
29. Grande place des Asphodèles
30. Place des crieurs
31. Caserne de la 3e Phalange
32. Grand jardin botanique
33. Froide-ville

District E.
34. Bibliothèque des merlots
35. Esplanade orientale
36. Manufacture des Gueux
37. Square de Derisse
38. Jardin des allées sans fin

District F.
39. Archives cartographiques
40. Grande place de la Pecherie
41. Usines de la Vrille
42. Hotel des ventes de Sabreck
43. Place de Malmort
44. Parc du port de la Baille
45. Entrepots de la Baille

District G.
46. Prison de Fort-Scellé
47. Casernes d'Yphyrion
48. Observatoire des Remparts
49. Vieux remparts
50. Jardin de Mai
51. Hotel de Geriond
52. Tribunal civil
53. Place du Corbeau
54. Square de la Mareille
55. Parc Vanlaste
56. Jardins des avenues
57. Manufacture Markeland
58. Usines des traverses
59. Capitainerie du Phleren

Stations du Gueulard
A. Gare la gueularde
B. Baille-tombe
C. Horologium
D. Estaban-Traverses
E. Général Sardeille
F. La Berne
G. Grand trou
H. Merian-Phader
I. Jardin botanique
J. Marle[illegible]rne

Index des ponts
P1. Pont des roués
P2. Pont Henyon
P3. Pont du sabre
P4. Pont du sabord
P5. Passerelle de la Tarse
P6. Passerelle du bassin
P7. Pont du canal
P8. Pont impérial
P9. Vieux pont des traverses
P10. Pont de l'île
P11. Pont des rebus
P12. Pont Sardeille
P13. Pont du Phleren
P14. Pont de la pecherie
P15. Pont du fort

Autorisation n° 13-22 Bureau des Censures - Ministère de la Propagande

3 Nov. 157 NC Emp. Imprimé par R. Gressac - Imprimeries des Compagnies, 13 rue du Grand Corbeau, Yphyrion Dis. G.

« The City of Yphyrion

*Maxime **Plasse***

I have always taken a keen interest in old maps and the technical and aesthetic modes that prevailed in their design through the ages. By creating the city of Yphyrion, I wanted to try to approach, using digital techniques, the style of urban maps from the late nineteenth century and their specific characteristics (top-down map view, crosshatched blocks of buildings, streets labels directly following the routes, etc.).

Yphyrion is a fictional city, but one whose structure is consistent enough to seem realistic. The integration of certain elements into the map and its legend (railways, underground metro system, factories etc.) clearly evokes a city of a newly industrialized, booming society

To highlight the city itself, and to give a consistent overall aesthetic, I decided to use a contrast of complementary colors. The green tones of the wooded and water areas, as well as more subtle ones at the edge of the map, serve to enhance the red tones used for the neighborhoods and buildings (and which also bring out some elements of text in the titles and the legend).

Linking past to present, fiction to reality: these were the objectives of this work.

Yphyrion is a fictional city, but one whose structure is consistent enough to seem realistic.

The J.B. Physical Map of the World »

Nat ***Case***

Of course I love old maps. Well, not all old maps, but I draw a lot of inspiration from a few examples: oil company road maps of the 1950s, Philadelphia streetcar maps of the late 1940s, and Hagstrom's maps from my childhood. I especially love the maps of Bartholomew and Sons of Edinburgh. Their blue-cover touring maps and their pre-digital *London Plan* have been touchstones for me since before I started making maps. The color palette of their older hypsometric (elevation) tint scheme (before they switched to CMYK printing), the delicacy of the engraved lettering…all have roots in the particular printing techniques they used, and are based in part on maintaining registration in a world where each ink color was printed on a separate pass through a one-color press.

This map is part of an ongoing exercise in reproducing the look and feel of early twentieth century maps by Bartholomew's. In this case I started with a map in a school atlas prepared for Oxford University Press in 1917. The map was called "bathy-orographical," a name they made up to describe showing the Earth's surface both above and below the ocean. Nowadays we'd just call it a "physical map" and leave it at that.

Almost all my data were taken from Natural Earth and used without prejudice. SRTM data were used to create the land hypsometric tints. I relied heavily on the font Sirenne, which to my mind is the best at emulating the standard engraved hand of a century ago. Much of the type in the Bartholomew's map was drawn along a curve; the difference in labor time between that and type on a straight line was negligible when every letter was hand drawn. As an exercise (in an era when it's substantially more work to put type on a curve), I placed all but a few labels that way. It's interesting to me what a difference it made in making this map evocative of that era.

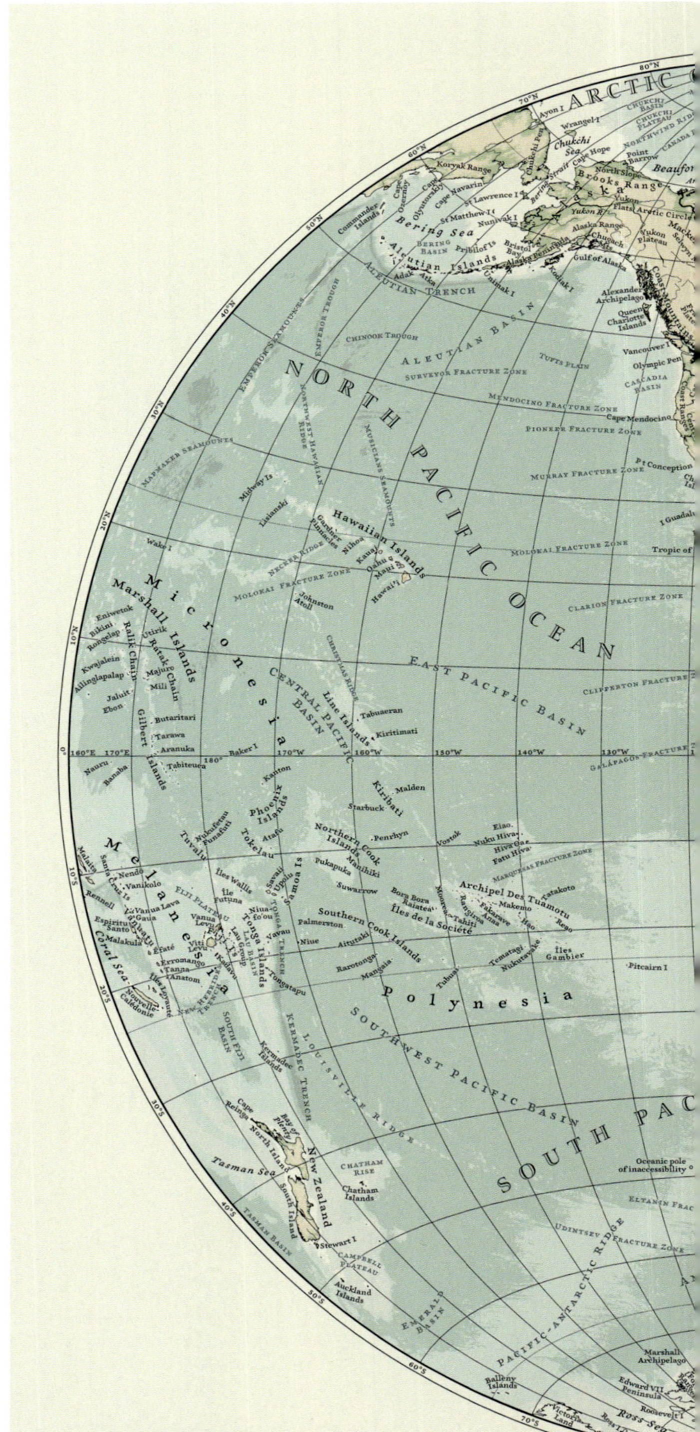

Lambert Azimuthal Equal-Area Projection.
Style based on the "Bathy-orographical" map of the world in the *Advanced Atlas of Physical and Political Geography*, published by Oxford University Press in 1917.

THE J.B. PHYSICAL MAP OF THE WORLD

In tribute to the cartography of John George Bartholomew.
This map created by Nat Case, INCase, LLC.

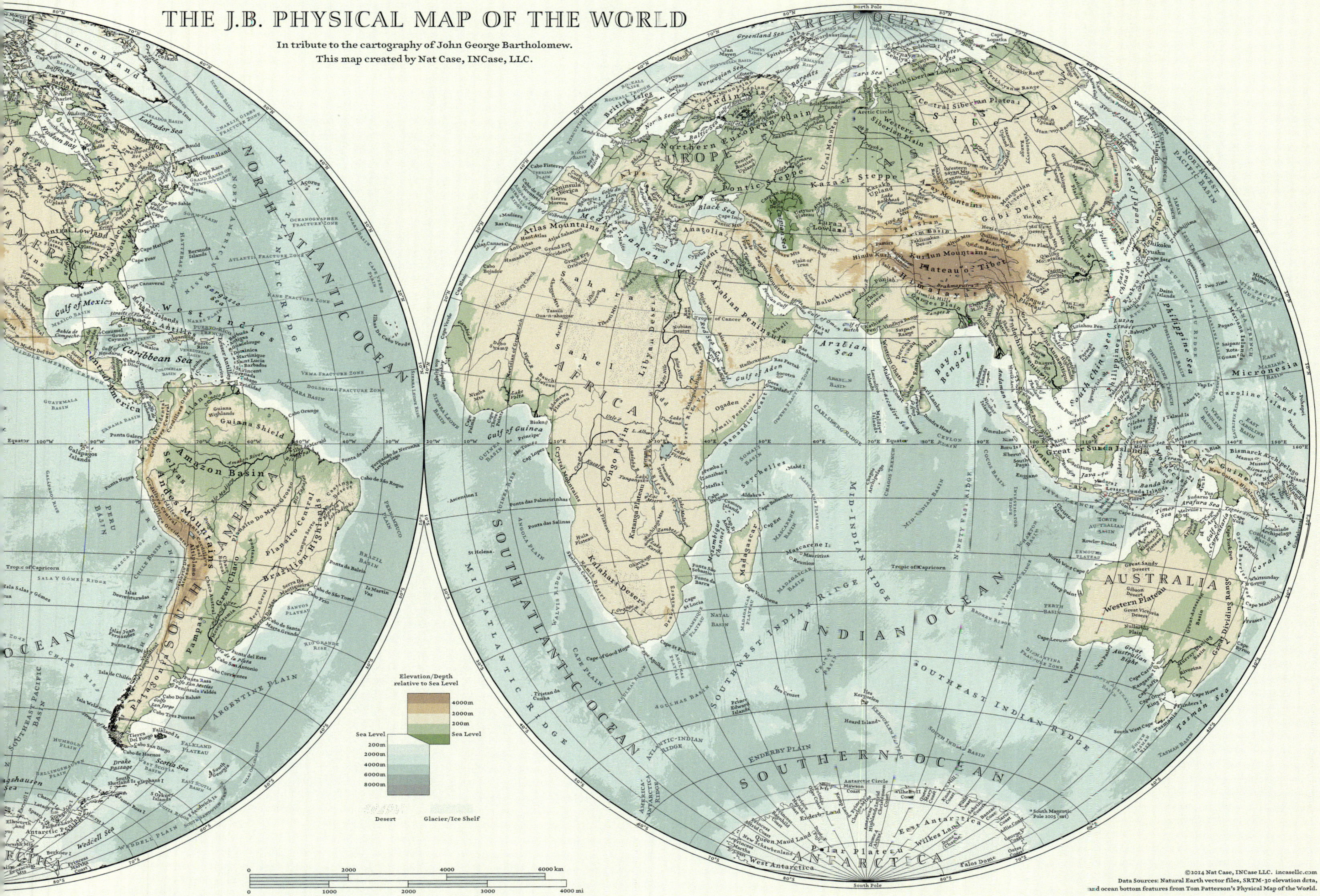

Data Sources: Natural Earth vector files, SRTM-30 elevation data, and ocean bottom features from Tom Patterson's Physical Map of the World.

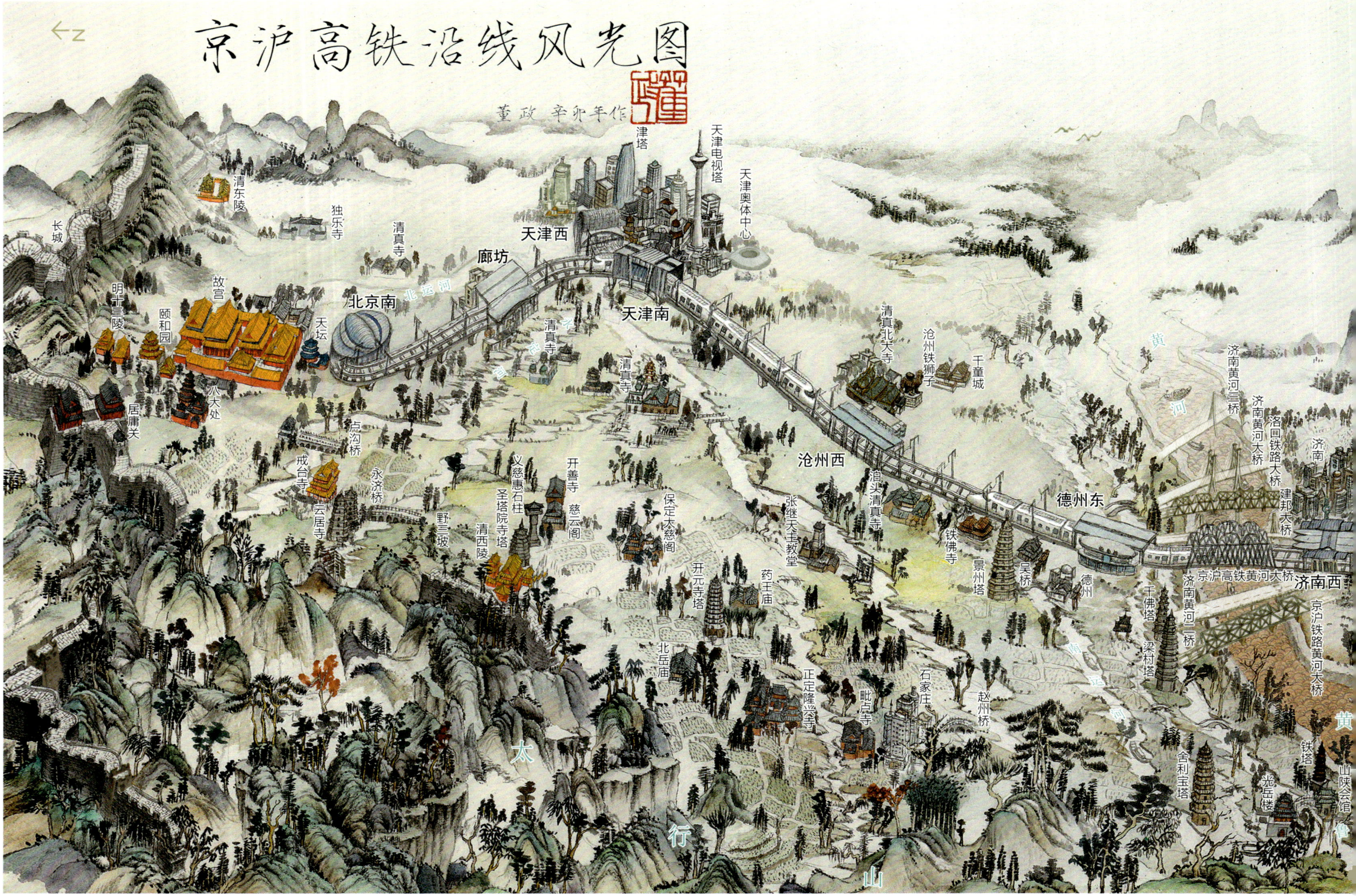

京沪高铁沿线风光图
董政 辛卯年作
津塔
天津电视塔
天津奥体中心
清东陵
独乐寺
清真寺
长城
天津西
廊坊
北京南
北运河
天津南
故宫
明十三陵
颐和园
天坛
永定河
清真寺
清真寺
居庸关
八大处
卢沟桥
戒台寺
永济桥
云居寺
野三坡
清西陵
圣塔院寺塔
义慈惠石柱
开善寺
慈云阁
保定大慈阁
开元寺塔
药王庙
北岳庙
正定隆兴寺
毗卢寺
石家庄
赵州桥
沧州西
张继天主教堂
泊头清真寺
铁佛寺
景州塔
吴桥
德州
德州东
清真北大寺
沧州铁狮子
千童城
南运河
黄河
济南黄河三桥
济南黄河大桥
洛口铁路大桥
建邦大桥
济南
京沪高铁黄河大桥
济南西
济南黄河二桥
千佛塔
梁村塔
京沪铁路黄河大桥
舍利宝塔
光岳楼
铁塔
山陕会馆
黄
鲁
太
行
山

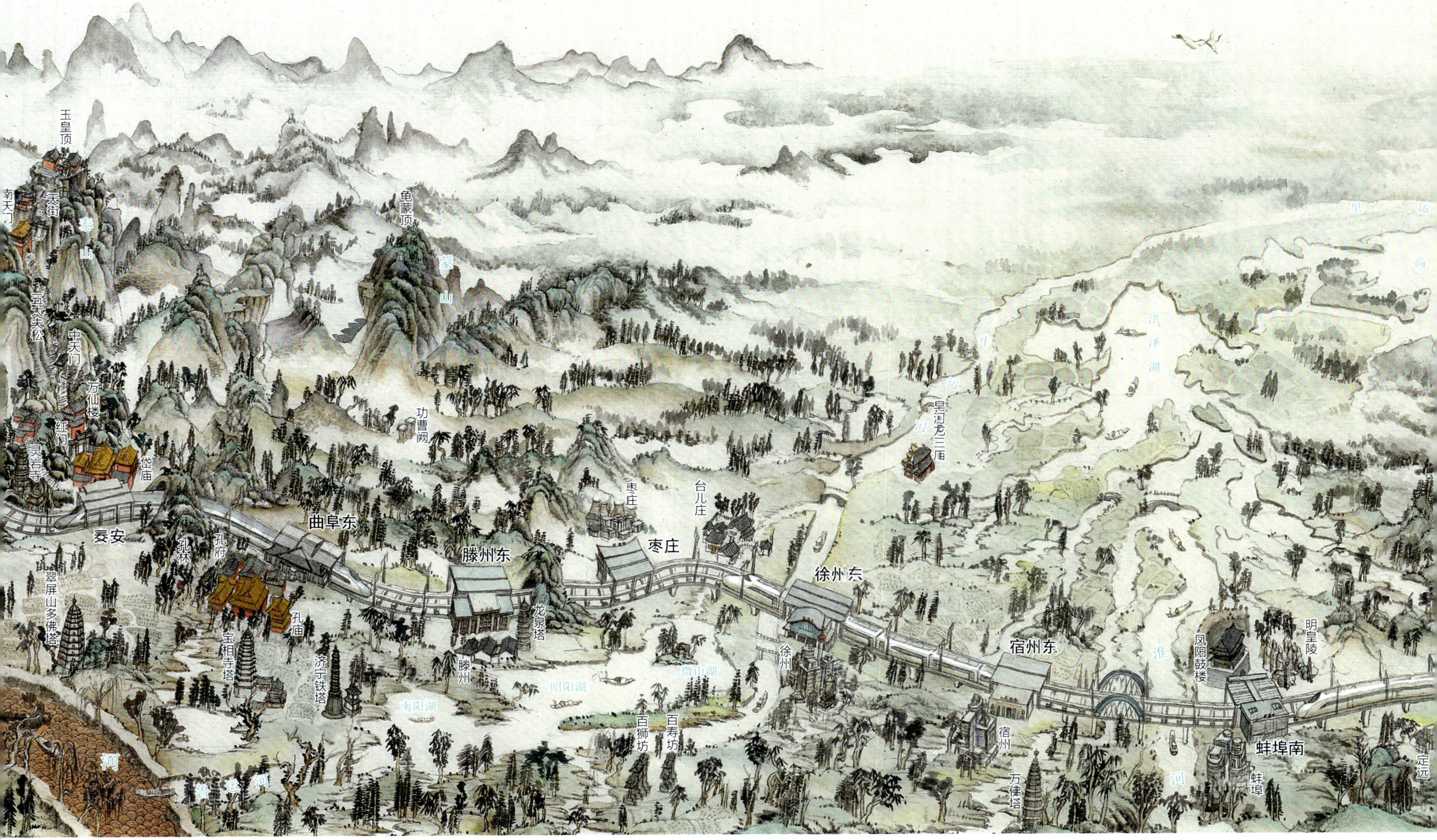

玉皇顶
南天门
天街
泰山
五大夫松
中天门
万仙楼
红门
灵岩寺
岱庙
泰安
翠屏山多佛塔
孔林
孔府
孔庙
宝相寺塔
曲阜东
龟蒙顶
蒙山
滕州东
龙泉塔
滕州
济宁铁塔
南阳湖
昭阳湖
微山湖
枣庄
台儿庄
百狮坊
百寿坊
徐州东
徐州
皂河龙王庙
洪泽湖
宿州东
宿州
凤阳鼓楼
明皇陵
淮河
蚌埠南
蚌埠
定远

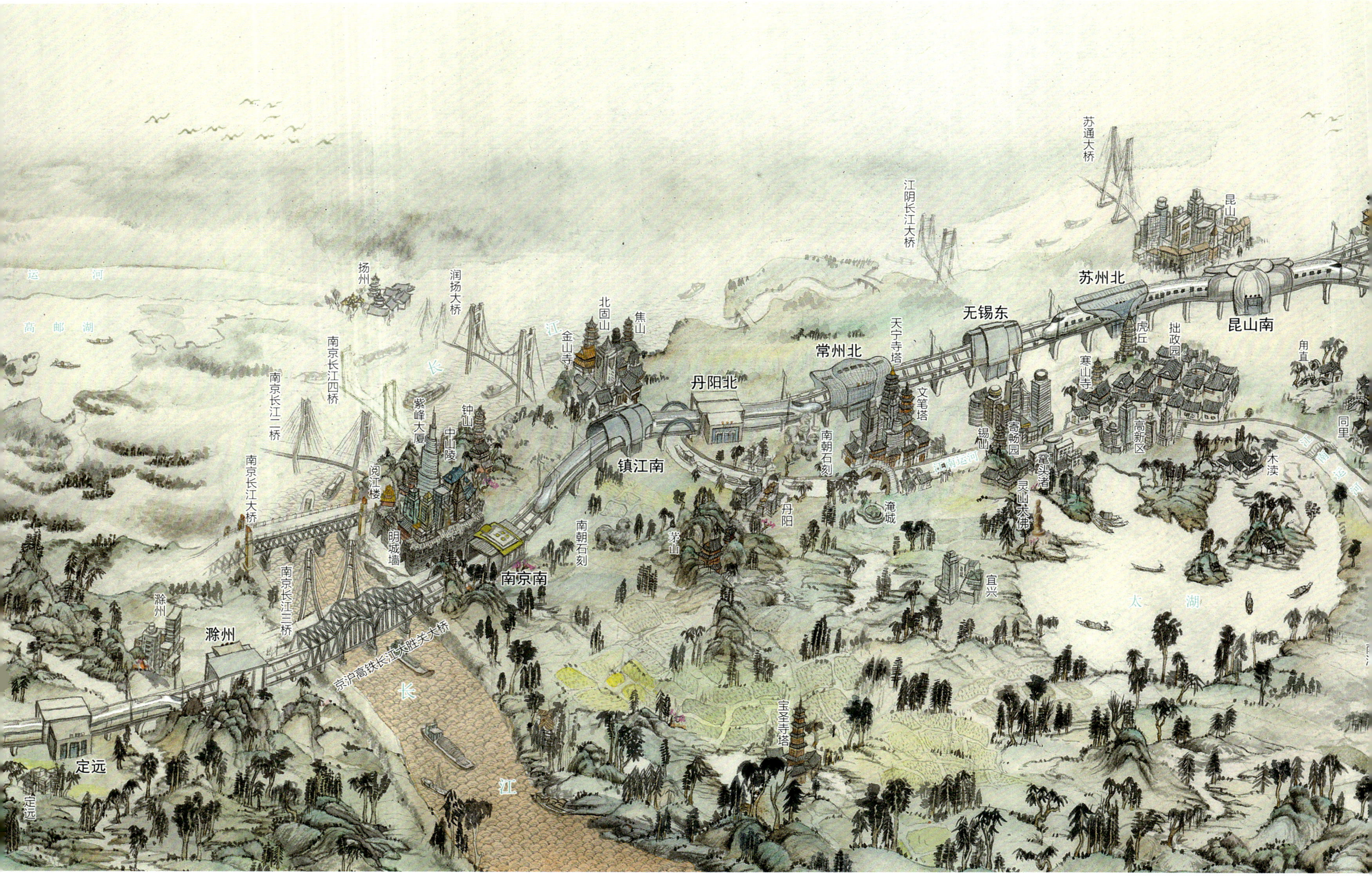

运河
高邮湖
扬州
润扬大桥
南京长江四桥
南京长江二桥
南京长江大桥
南京长江三桥
京沪高铁长江大胜关大桥
滁州
滁州
定远
定远
长
江
长
江
阅江楼
明城墙
紫峰大厦
中山陵
钟山
南京南
金山寺
北固山
焦山
镇江南
南朝石刻
南朝石刻
茅山
丹阳北
丹阳
常州北
天宁寺塔
文笔塔
淹城
宝圣寺塔
江阴长江大桥
无锡东
锡山
寄畅园
鼋头渚
灵山大佛
宜兴
太湖
苏通大桥
苏州北
虎丘
寒山寺
拙政园
高新区
木渎
昆山
昆山南
甪直
同里

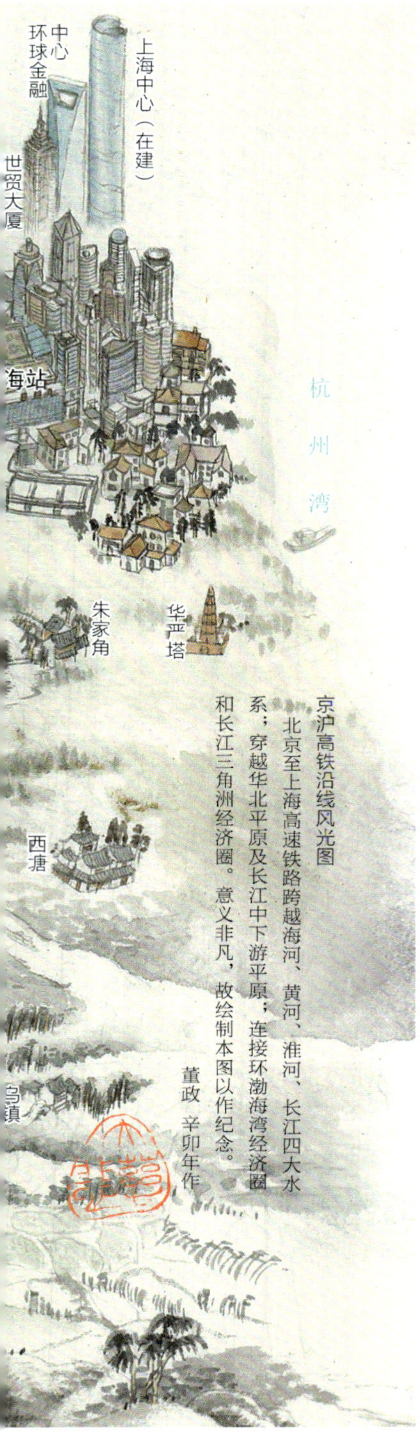

《京沪高铁沿线风光图
Map of the Beijing-Shanghai High-Speed Railway

董政 *Dong Zheng*

Commentary translated by Xu Wang-Angsüsser

The outstanding maps from pre-modern China were intended, first and foremost, to be great paintings. They were more high achievements in art, pleasing to people's eyes, than just efficient tools for practical use. One example, the *Chart of the Yellow River* (early Qing Dynasty, circa 1700), was an extraordinary work of art submitted to the Kangxi Emperor (1662–1722), and was also one inspiration for my map.

In particular, the overall perspective follows this masterpiece. Instead of the Yellow River, my map extends along the Beijing-Shanghai high-speed railway, which began service in 2011, connecting the two most important cities in mainland China faster than ever. I tried to follow my predecessors in drawing a map which informs about landscapes and culture, while giving an enjoyable high-speed feeling. Using a bird's-eye view, I wanted to give the impression of how, as this modern high-tech transportation carries people from one city to another, it also connects historical heritage and beautiful natural scenery.

Nowadays in our modern society, simplification and economization are advocated. Practical value is considered more than aesthetic needs. But we should take the time needed to make and enjoy beautiful things. In my opinion the greatest artists first acquire excellent painting techniques, and then they can create their aesthetic works in a simplified style. In the *Map of the Beijing-Shanghai High-Speed Railway* I tried to merge outstanding historic buildings with modern ones in a harmonious way, thus presenting the whole scenery in its most beautiful form. In a rapid-development environment, we should not forget to preserve tradition and beauty.

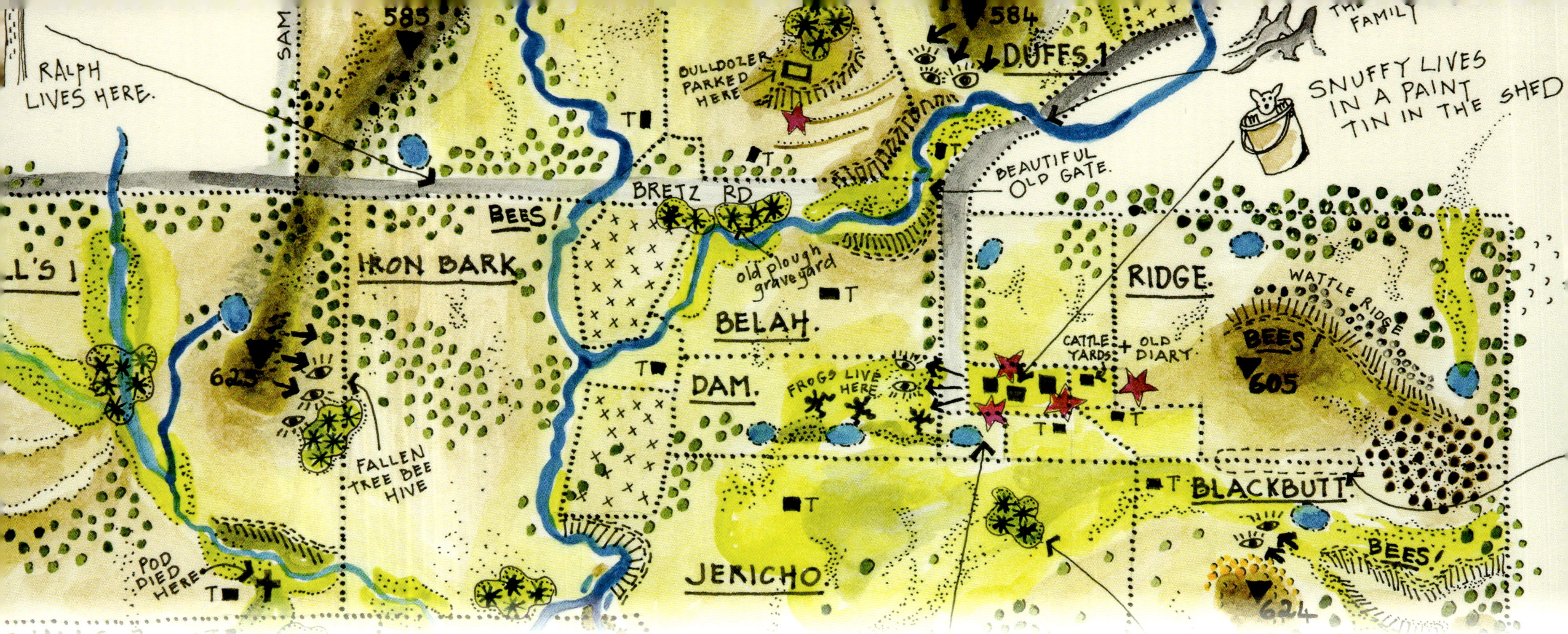

The Family Farm »

Alex **Hotchin**

On a recent visit home to see my family, I decided to map the family farm. It's a place quite new to me and my family, but after all the exploring, recording, and drawing, I feel like we've gotten to know each other.

The abundance of textures, lines, intersections, borders, and shapes weave together like a patterned, hand-woven blanket. I'm interested in the pathways traversing over and around nature's obstacles, the pattern a collection of trees make, and how a line makes a border which then creates a shape. To understand these patterns, I needed to draw them. In this map the patterns of the landscape are overlaid with significant personal stories, like where Buzz, the family dog, is buried, where a family of water dragons lives, where the beehives stand, and which part of the airspace the eagles like.

This map tells a story of a place that pictures and words could not. It pays respect to all the things that make a place special: the plants and animals that call this place home, and events that have occurred here that make this place home to our family.

Over several weeks, I walked and rode around Jericho observing and recording topographical features. Conversations with my father revealed his memories and experiences; too many stories to include here, but they were told and listened to as part of the exercise of drawing this map. I learned much about him and about how important a connection with land is to our sense of place. This taught me the lesson that a drawn map can teach much to the observer, but the maker also learns much.

This map was created with a brush and ink, and then overlaid with pen and ink. It was gifted to my father as a way for me to tell him how beautiful I consider Jericho to be, and how joyful we feel when we visit the family farm!

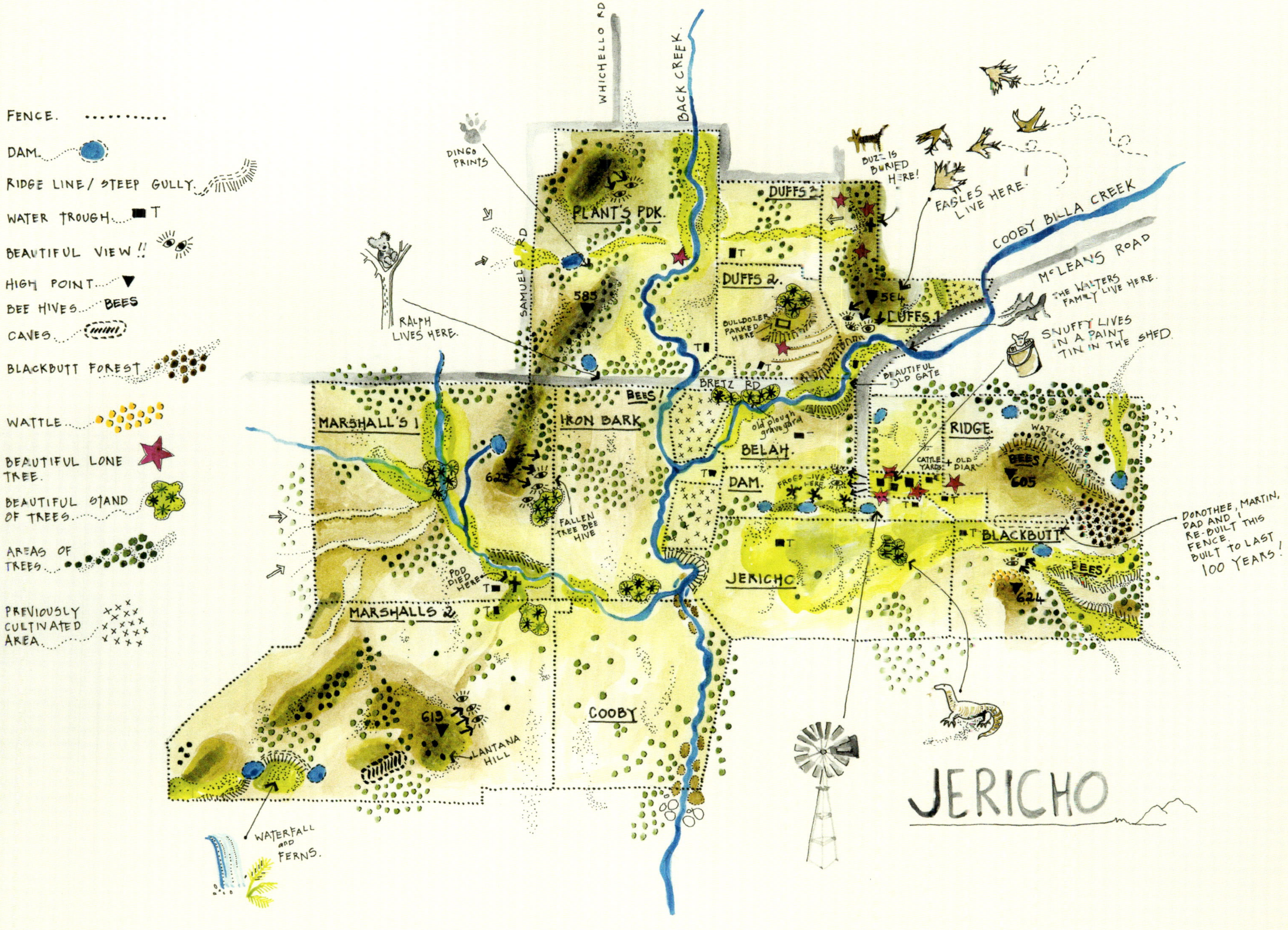

FENCE.
DAM.
RIDGE LINE / STEEP GULLY.
WATER TROUGH.
BEAUTIFUL VIEW!!
HIGH POINT.
BEE HIVES.
BEES
CAVES.
BLACKBUTT FOREST
WATTLE
BEAUTIFUL LONE TREE.
BEAUTIFUL STAND OF TREES.
AREAS OF TREES.
PREVIOUSLY CULTIVATED AREA.
WHICHELLO RD
BACK CREEK
DINGO PRINTS
SAMUELS RD
PLANT'S PDK.
585
RALPH LIVES HERE.
BUZ IS BURIED HERE!
EAGLES LIVE HERE.
DUFFS 2
DUFFS 2.
DUFFS 1
584
BULLDOZER PARKED HERE
COOBY BILLA CREEK
McLEANS ROAD
THE WALTERS FAMILY LIVE HERE.
SNUFFY LIVES IN A PAINT TIN IN THE SHED.
BEAUTIFUL OLD GATE
BRETZ RD
old plough graveyard
MARSHALL'S 1
IRON BARK
BEES
BELAH
RIDGE
WATTLE RIDGE
BEES
605
CATTLE YARDS
OLD DIAR
DAM.
FROGS LIVE HERE
623
FALLEN TREE BEE HIVE
DOROTHEE, MARTIN, DAD AND I RE-BUILT THIS FENCE. BUILT TO LAST 100 YEARS!
BLACKBUTT
BEES
624
JERICHO
POD DIED HERE
MARSHALLS 2
613
LANTANA HILL
COOBY
WATERFALL and FERNS.
JERICHO

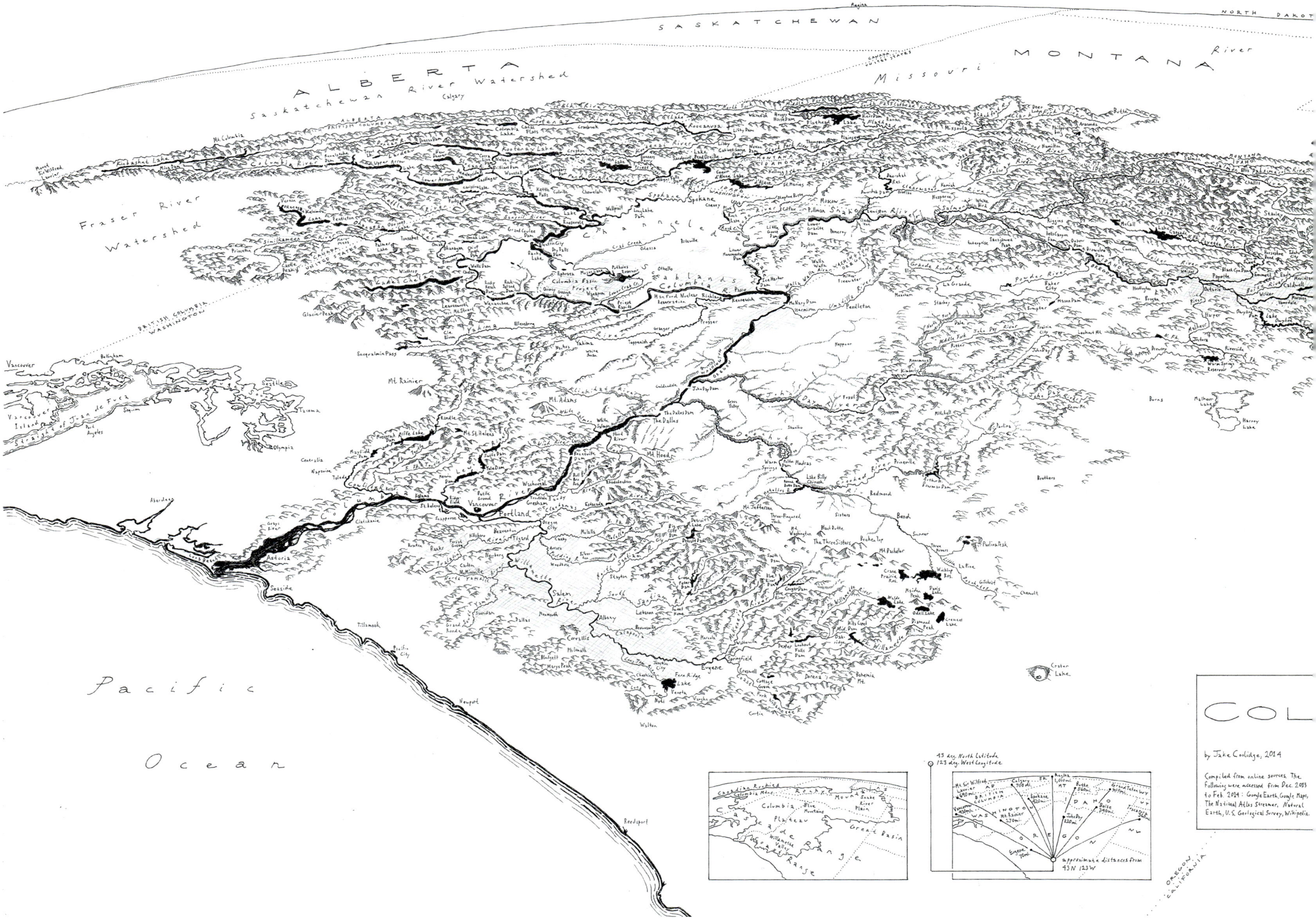

SASKATCHEWAN
ALBERTA
Saskatchewan River Watershed
Calgary
MONTANA
Missouri River
NORTH DAKOTA
Fraser River Watershed
BRITISH COLUMBIA
WASHINGTON
Spokane
Channeled Scablands
Columbia Basin Project
Columbia River
Snake River
Yakima
Mt. Rainier
Mt. Adams
Mt. St. Helens
Mt. Hood
The Dalles
Vancouver
Portland
Astoria
Seaside
Salem
Albany
Corvallis
Eugene
Springfield
Bend
Crater Lake
Tillamook
Newport
Reedsport
Pacific Ocean
Straight of Juan de Fuca
Vancouver Island
Seattle
Tacoma
Olympia
Pendleton
Boise
Butte
Flathead Lake
Kootenay Lake
Willamette River
Deschutes River
John Day River
OREGON
CALIFORNIA
COL
by Jake Coolidge, 2014
Compiled from online sources. The following were accessed from Dec. 2013 to Feb. 2014: Google Earth, Google Maps, The National Atlas Streamer, Natural Earth, U.S. Geological Survey, Wikipedia.
Columbia Plateau
Cascade Range
Coast Range
Willamette Valley
Great Basin
Snake River Plain
Rocky Mountains
43 deg. North Latitude
123 deg. West Longitude
approximate distances from 43N 123W

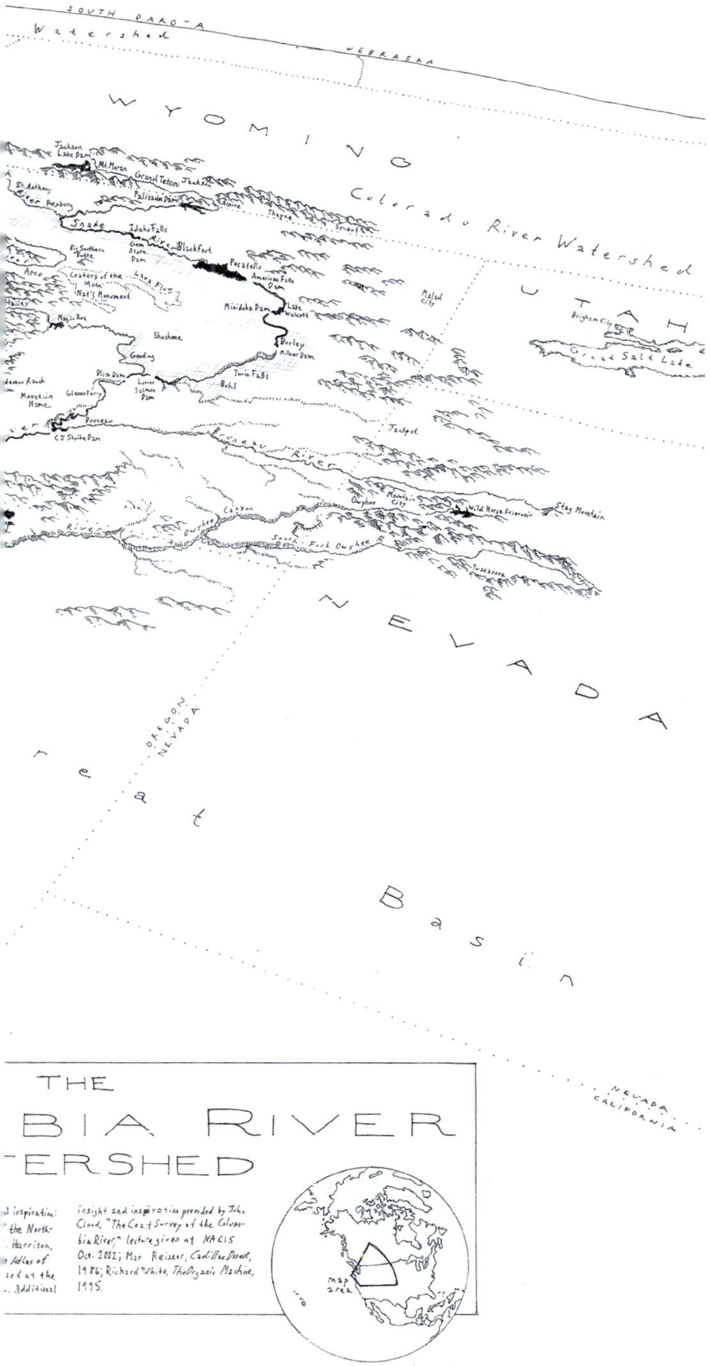

« The Columbia River Watershed

Jake ***Coolidge***

Watersheds encompass many streams whose fates are linked together, combining to form complex networks across a region bound by topography and gravity. By the time the Columbia River empties into the Pacific Ocean, it has drawn together water from tributaries originating in seven US states and one Canadian province. As the streams of this region find a common outlet to the Pacific, this river system has also formed the basis for the social, cultural, economic, and political linkages that comprise much of the Pacific Northwest, the region in which I was born and raised.

I wanted to create a map of the Columbia River watershed that at once highlights the interconnected nature of the river system while also illuminating the tremendous diversity of terrain and places found within the region. Emphasis is placed on the watershed by omitting most features outside of its boundary. The globe-perspective view was selected to emphasize the great reach of the watershed and reframe the viewer's perspective, helping to cast a familiar place in a new light.

I adopted a globe-view perspective after returning to the work of Richard Edes Harrison (1901–1994), particularly his maps of the Second World War published in *Fortune* magazine in the 1940s. Harrison's work stimulated renewed interest in geography for a general audience, with maps that expressed global interconnectedness and movement across landscapes. The terrain representations of Erwin Raisz (1893–1968) were also a significant influence, and his masterful map, *Landforms of the Northwestern States*, portrays rivers in solid black ink, allowing them to visually come to the fore. I chose to retain this symbolization in my own depiction.

My broader aim in compiling and drawing maps by hand is to better understand and maintain pre-digital cartographic techniques. Drawing complex geographies by hand not only gives me a much deeper understanding of the places I am depicting; the process also informs all aspects of my cartographic practice by making the process of selection, abstraction, generalization, and symbolization a tangible, lived experience.

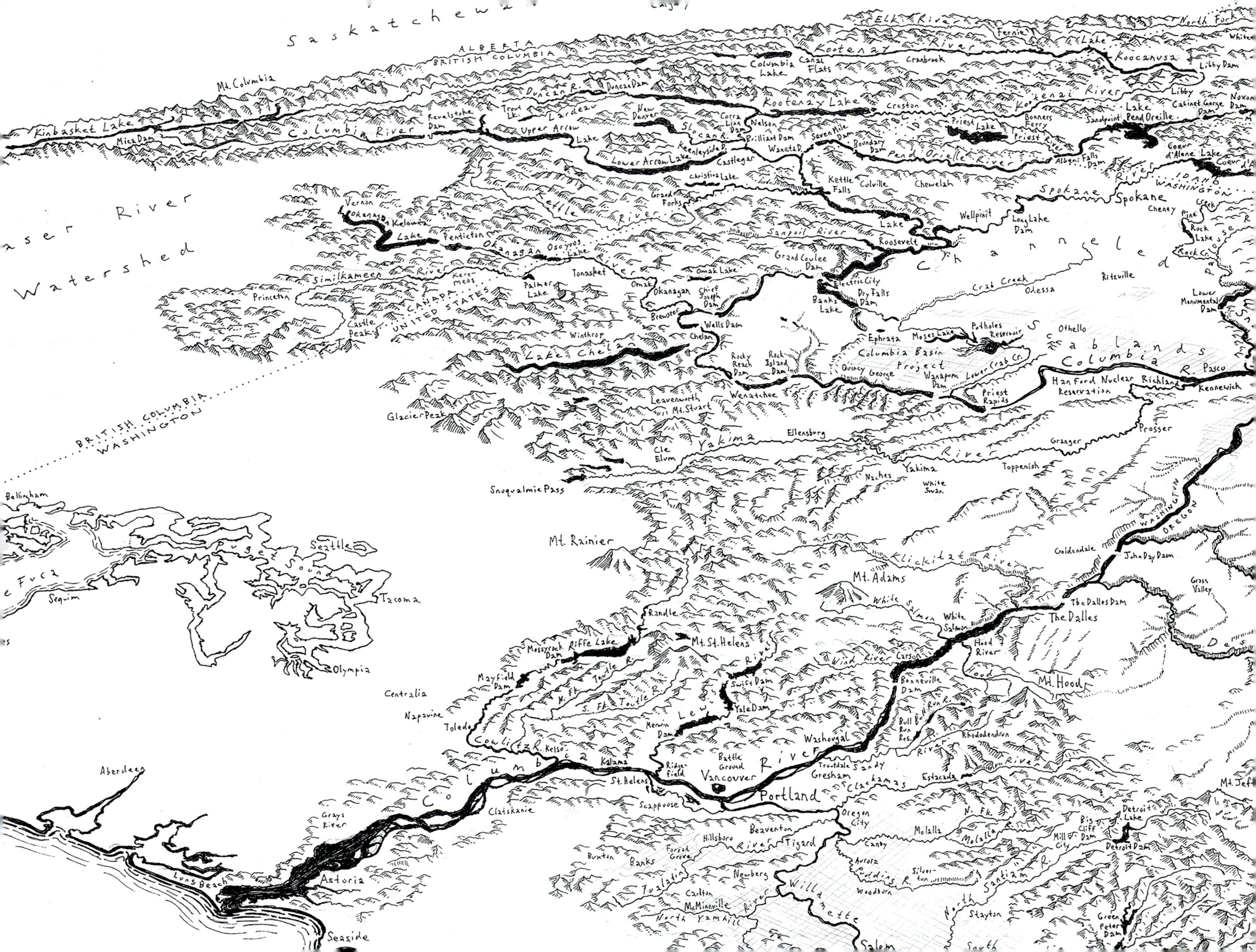

Saskatchewa
ALBERTA
BRITISH COLUMBIA
Elk River
North Fork
Fernie
Kootenay River
Lake Koocanusa
Cranbrook
Columbia Lake
Canal Flats
Libby Dam
Mt. Columbia
Duncan R.
Duncan Dam
Kootenay Lake
Creston
Kootenai River
Libby
Cabinet Gorge Dam
Noxon
Kinbasket Lake
Mica Dam
Columbia River
Revelstoke Dam
Trout Lk.
Lardeau
New Denver
Slocan R.
Corra Linn Dam
Nelson
Brilliant Dam
Seven Mile Dam
Waneta D.
Boundary Dam
Priest Lake
Bonners Ferry
Priest River
Sandpoint
Lake Pend Oreille
Upper Arrow Lake
Lower Arrow Lake
Keenleyside D.
Castlegar
Pend Oreille River
Albeni Falls Dam
Coeur d'Alene Lake
Christina Lake
Kettle Falls
Colville
Chewelah
IDAHO
WASHINGTON
Spokane River
Spokane
Cheney
Fraser River Watershed
Vernon
Okanagan Lake
Kelowna
Penticton
Okanagan River
Osoyoos Lake
Kettle River
Grand Forks
Sanpoil River
Lake Roosevelt
Wellpinit
Long Lake Dam
Rock Lake
Rock Cr.
Palouse R.
Channeled Scablands
Grand Coulee Dam
Electric City
Dry Falls Dam
Banks Lake
Crab Creek
Odessa
Ritzville
Lower Monumental Dam
Similkameen River
Keremeos
Tonasket
Omak
Omak Lake
Palmer Lake
Princeton
CANADA
UNITED STATES
Okanagan
Chief Joseph Dam
Brewster
Castle Peak
Wells Dam
Winthrop
Chelan
Lake Chelan
Ephrata
Moses Lake
Potholes Reservoir
Othello
Columbia Basin Project
Rocky Reach Dam
Rock Island Dam
Quincy
George
Wanapum Dam
Lower Crab Cr.
Columbia R.
Pasco
Hanford Nuclear Reservation
Richland
Kennewick
Priest Rapids
Wenatchee
Leavenworth
Mt. Stuart
Glacier Peak
BRITISH COLUMBIA
WASHINGTON
Ellensburg
Yakima River
Cle Elum
Prosser
Granger
Yakima
Naches
Toppenish
White Swan
Snoqualmie Pass
Bellingham
Puget Sound
Seattle
Fuca
Sequim
Tacoma
Olympia
Mt. Rainier
Klickitat River
Goldendale
John Day Dam
WASHINGTON
OREGON
Mt. Adams
White Salmon
Grass Valley
The Dalles Dam
The Dalles
Randle
Mossyrock Dam
Riffe Lake
Mt. St. Helens
Wind River
Carson
Hood River
Mayfield Dam
Toutle R.
N. Fk. Toutle R.
S. Fk. Toutle R.
Swift Dam
Yale Dam
Lewis River
Bonneville Dam
Mt. Hood
Centralia
Napavine
Toledo
Merwin Dam
Bull Run Res.
Bull Run R.
Rhododendron
Cowlitz R.
Kelso
Kalama
Washougal
Columbia River
Ridgefield
Battle Ground
Vancouver
Troutdale
Sandy River
Gresham
Estacada
Aberdeen
St. Helens
Portland
Clackamas
Scappoose
Grays River
Clatskanie
Oregon City
Beaverton
Tigard
Hillsboro
Molalla
N. Fk. Molalla
Detroit Lake
Big Cliff Dam
Mill City
Detroit Dam
Mt. Jeff
Buxton
Banks
Forest Grove
Canby
Aurora
Pudding R.
Silverton
Santiam R.
Long Beach
Astoria
Tualatin River
Carlton
Newberg
McMinnville
Willamette River
Woodburn
North Santiam
Stayton
North Yamhill River
Seaside
Salem
Green Peter Dam

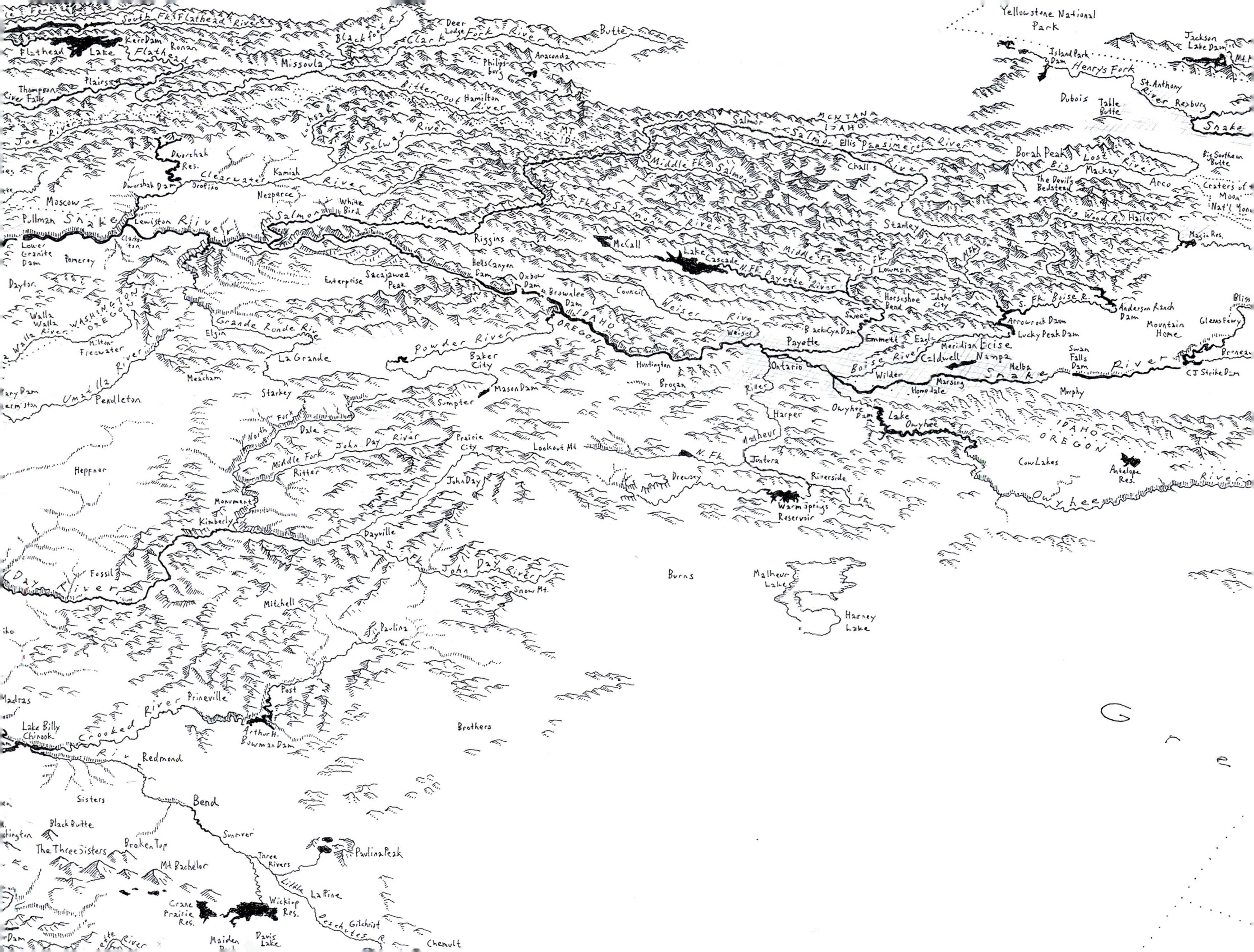

South Fk. Flathead River
Kerr Dam
Ronan
Flathead Lake
Flathead
Blackfoot R.
Deer Lodge
Clark Fork River
Butte
Missoula
Philipsburg
Anaconda
Plains
Thompson River Falls
Bitterroot River
Hamilton
Lochsa R.
Selway River
Joe River
Dworshak Res.
Dworshak Dam
Orofino
Clearwater
Kamiah
River
Moscow
Pullman
Snake River
Lewiston
Clarkston
Nezperce
White Bird
Salmon River
Lower Granite Dam
Pomeroy
Dayton
Walla Walla River
WASHINGTON
OREGON
Milton-Freewater
Riggins
Hells Canyon Dam
Oxbow Dam
Brownlee Dam
Enterprise
Sacajawea Peak
Grande Ronde River
Elgin
La Grande
Powder River
Baker City
Mason Dam
Meacham
Umatilla River
Pendleton
Starkey
Sumpter
North Fork
Dale
John Day River
Middle Fork
Ritter
Prairie City
John Day
Heppner
Monument
Kimberly
Dayville
S. Fk. John Day River
Fossil
Day River
Snow Mt.
Mitchell
Paulina
Madras
Prineville
Crooked River
Post
Lake Billy Chinook
Arthur R. Bowman Dam
Redmond
Brothers
Sisters
Bend
Black Butte
The Three Sisters
Broken Top
Sunriver
Mt Bachelor
Three Rivers
Paulina Peak
Little
La Pine
Crane Prairie Res.
Wickiup Res.
Gilchrist
Deschutes R.
Davis Lake
Maiden
Chemult
Yellowstone National Park
Jackson Lake Dam
Island Park Dam
Henrys Fork
St. Anthony
Rexburg
Dubois
Table Butte
Snake
MONTANA
IDAHO
MT
ID
Salmon
Ellis
Pahsimeroi River
Challis
Middle Fk. Salmon R.
Borah Peak
Big Lost River
Mackay
Big Southern Butte
Arco
Craters of the Moon Nat'l Mon.
The Devil's Bedstead
Big Wood R.
Hailey
Magic Res.
S. Fk. Salmon River
McCall
Lake Cascade
N. Fk. Payette River
Middle Fk.
Stanley
Lowman
Council
IDAHO
OREGON
Weiser River
Weiser
Horseshoe Bend
Idaho City
S. Fk. Boise R.
Anderson Ranch Dam
Black Cyn Dam
Sweet
Arrowrock Dam
Lucky Peak Dam
Bliss
Glenns Ferry
Mountain Home
Payette
Emmett
Eagle
Meridian
Boise
Caldwell
Nampa
Boise River
Huntington
Ontario
Wilder
Melba
Swan Falls Dam
C J Strike Dam
Bruneau
Brogan
Marsing
Homedale
Murphy
Harper
Owyhee Dam
Lake Owyhee
Lookout Mt
Malheur
N. Fk.
Juntura
Drewsey
Riverside
S. Fk.
Warm Springs Reservoir
Cow Lakes
Antelope Res.
Owyhee River
Burns
Malheur Lake
Harney Lake

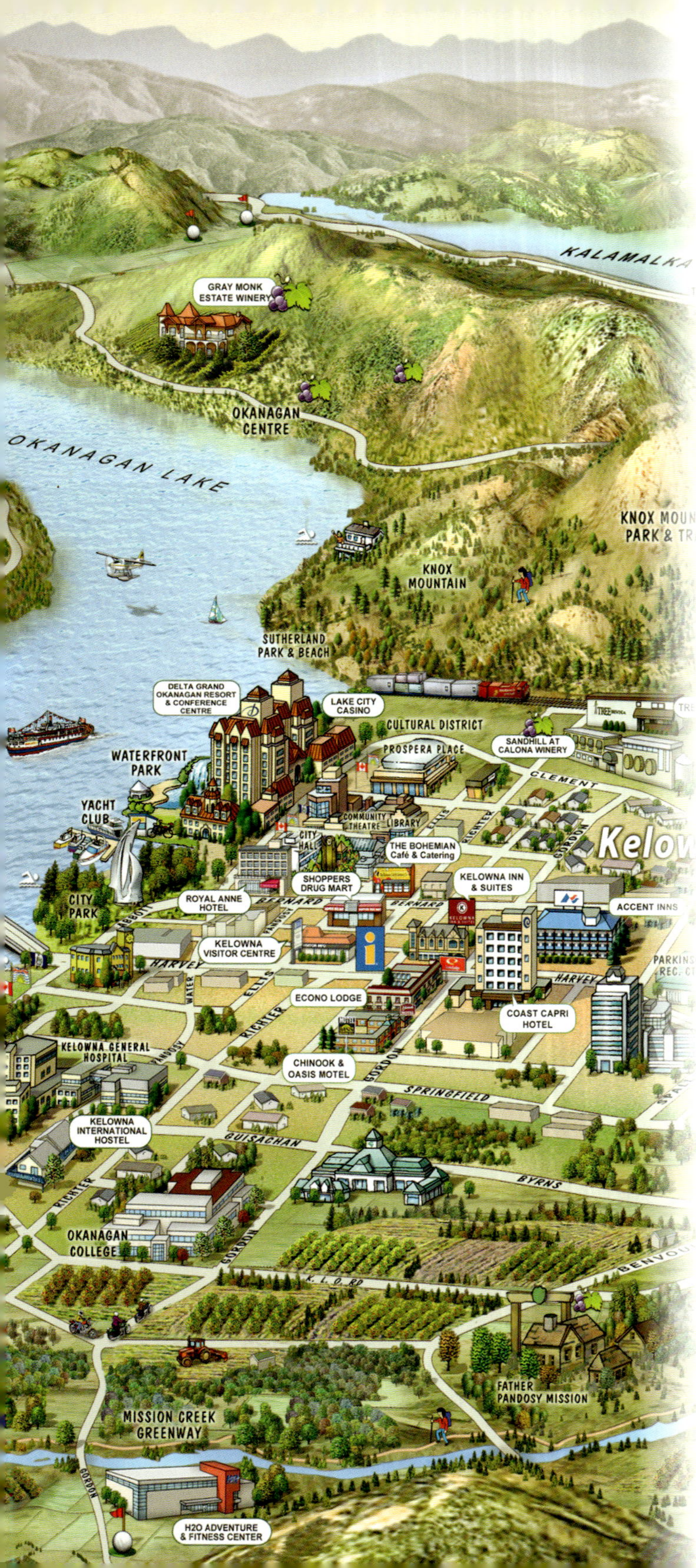

Kelowna, British Columbia »

Jean-Louis ***Rheault***

The mission here was to promote the tourism assets of the city of Kelowna in the heart of British Columbia's beautiful Okanagan Valley.

For tourism maps, the central challenge is always the same: how do you present the large-scale natural beauty of the overall area, along with the specific and much smaller man-made attractions that dot the landscape, and say it all in one image? My preferred solution is the pictorial oblique bird's-eye view approach with multiple interlocking scales.

The first stage is to create an overall concept of the area with a good old-fashioned pencil. I always try to do this by memory, so that I retain my impressions of a place rather than get overly influenced by what scale road maps tell us. As we all experience, our impression of 100 yards in a parking lot is a much greater distance than 300 yards between the cottage and the lake.

Once the main roads and individual buildings have been positioned temporarily as blocks, the landscape is painted with Photoshop brushes. The Okanagan Valley is an oasis in the middle of the vast, rugged Monashee Mountains. The color palette was chosen to render the lush greenery of this fertile, fruit-growing terrain watered by large attractive lakes and framed by arid, curvy, low mountains and sculpted hills.

After the landscape has risen, it's then back to the specific buildings, parks, orchards, etc., which are rendered in great detail. Rooftops are minimized and facades enhanced to conserve the street-view impression. Then lastly, the road network is built to connect and harmonize the differing scales of the manmade with its natural panoramic setting and, of course, reliably get you from A to B.

The result, which is displayed on large outdoor information signs, is that visitors can simply let their eyes roam around the map and absorb a comprehensive sense of the area without being distracted by insets, zooms, and legends. They get an organic experience of place, and that will guide and simplify any need for precise information on their smartphones.

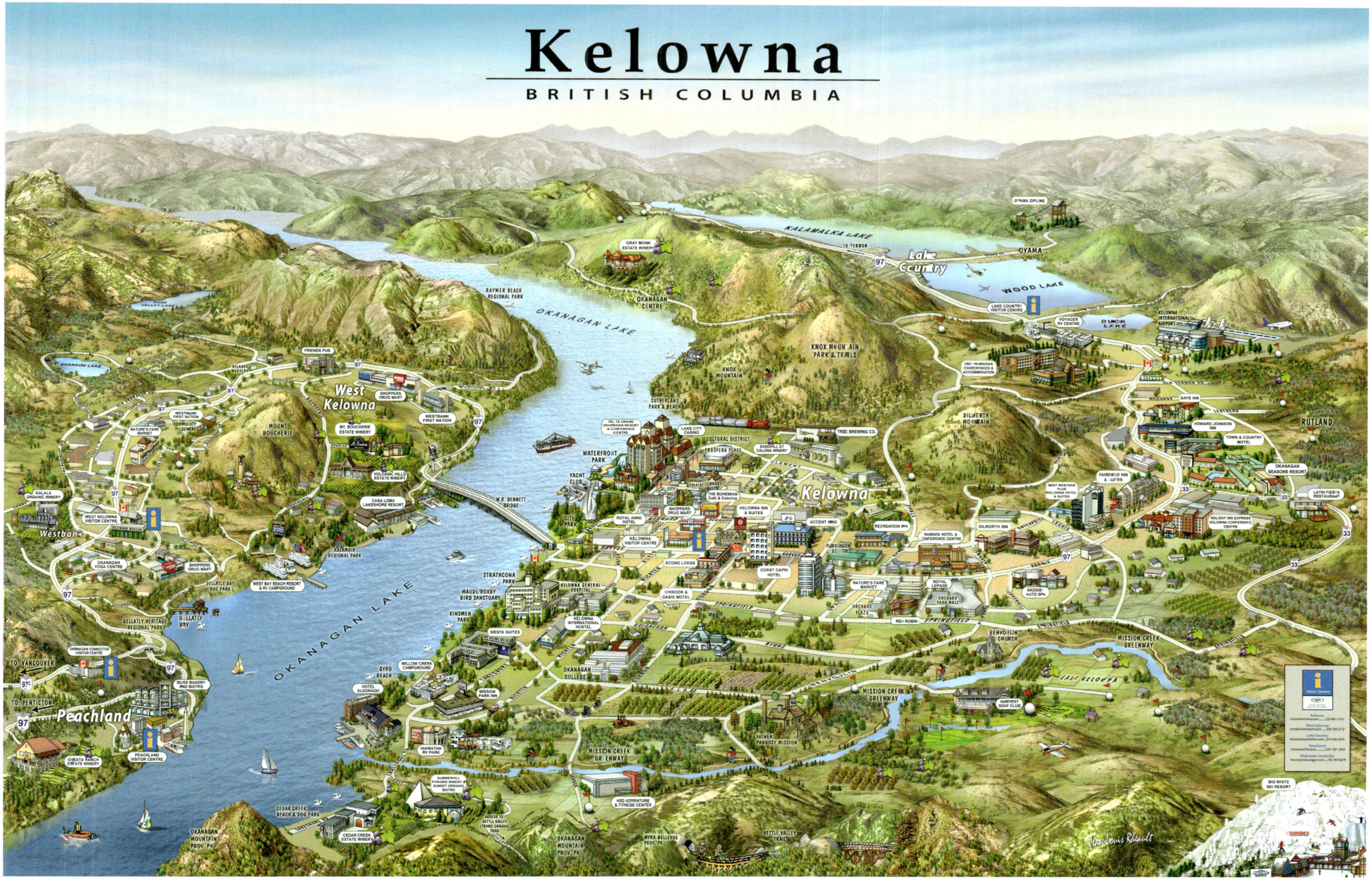

Kelowna
BRITISH COLUMBIA
KALAMALKA LAKE
TO VERNON
OYAMA
Lake Country
WOOD LAKE
DUCK LAKE
LAKE COUNTRY VISITOR CENTRE
KELOWNA INTERNATIONAL AIRPORT
GRAY MONK ESTATE WINERY
RAYMER BEACH REGIONAL PARK
OKANAGAN LAKE
OKANAGAN CENTRE
KNOX MOUNTAIN PARK & TRAILS
KNOX MOUNTAIN
SHANNON LAKE
FRIENDS PUB
West Kelowna
SHOPPERS DRUG MART
WESTBANK FIRST NATION
MOUNT BOUCHERIE
MT. BOUCHERIE ESTATE WINERY
VOLCANIC HILLS ESTATE WINERY
CASA LOMA LAKESHORE RESORT
KALALA ORGANIC WINERY
WEST KELOWNA VISITOR CENTRE
Westbank
OKANAGAN YOGA CENTRE
W.R. BENNETT BRIDGE
SUTHERLAND PARK & BEACH
WATERFRONT PARK
YACHT CLUB
CITY PARK
LAKE CITY CASINO
CULTURAL DISTRICT
PROSPERA PLACE
SANDHILL AT CALONA WINERY
TREE BREWING CO.
DILWORTH MOUNTAIN
RUTLAND
Kelowna
ROYAL ANNE HOTEL
KELOWNA VISITOR CENTRE
SHOPPERS DRUG MART
ACCENT INNS
RECREATION SPA
RAMADA HOTEL & CONFERENCE CENTRE
DILWORTH INN
FAIRFIELD INN & SUITES
HOLIDAY INN EXPRESS KELOWNA CONFERENCE CENTRE
OKANAGAN SEASONS RESORT
LATIN FIESTA RESTAURANT
TOWN & COUNTRY MOTEL
HOWARD JOHNSON INN
DAYS INN
ECONO LODGE
COAST CAPRI HOTEL
STRATHCONA PARK
KELOWNA GENERAL HOSPITAL
CHINOOK & OASIS MOTEL
NATURE'S FARE MARKET
ORCHARD PARK MALL
SKOGIE AUTO SPA
ROYAL LEPAGE
BENVOULIN CHURCH
MISSION CREEK GREENWAY
EAST KELOWNA
MAUDE/ROXBY BIRD SANCTUARY
KINSMEN PARK
SIESTA SUITES
KELOWNA INTERNATIONAL HOSTEL
OKANAGAN COLLEGE
GYRO BEACH
WILLOW CREEK CAMPGROUND
HOTEL ELDORADO
MISSION PARK INN
HIAWATHA RV PARK
FATHER PANDOSY MISSION
HARVEST GOLF CLUB
H2O ADVENTURE & FITNESS CENTER
BIG WHITE SKI RESORT
CEDAR CREEK BEACH & DOG PARK
CEDAR CREEK ESTATE WINERY
OKANAGAN MOUNTAIN PROV. PK
MYRA BELLEVUE PROV. PK
KETTLE VALLEY TRAIL
WEST BAY BEACH RESORT & RV CAMPGROUND
GELLATLY BAY DOG PARK
GELLATLY HERITAGE REGIONAL PARK
GELLATLY BAY
OKANAGAN CONNECTOR VISITOR CENTRE
TO VANCOUVER
TO PENTICTON
Peachland
BLISS BAKERY AND BISTRO
GORAN RANCH ESTATE WINERY
PEACHLAND VISITOR CENTRE
97
97C
33

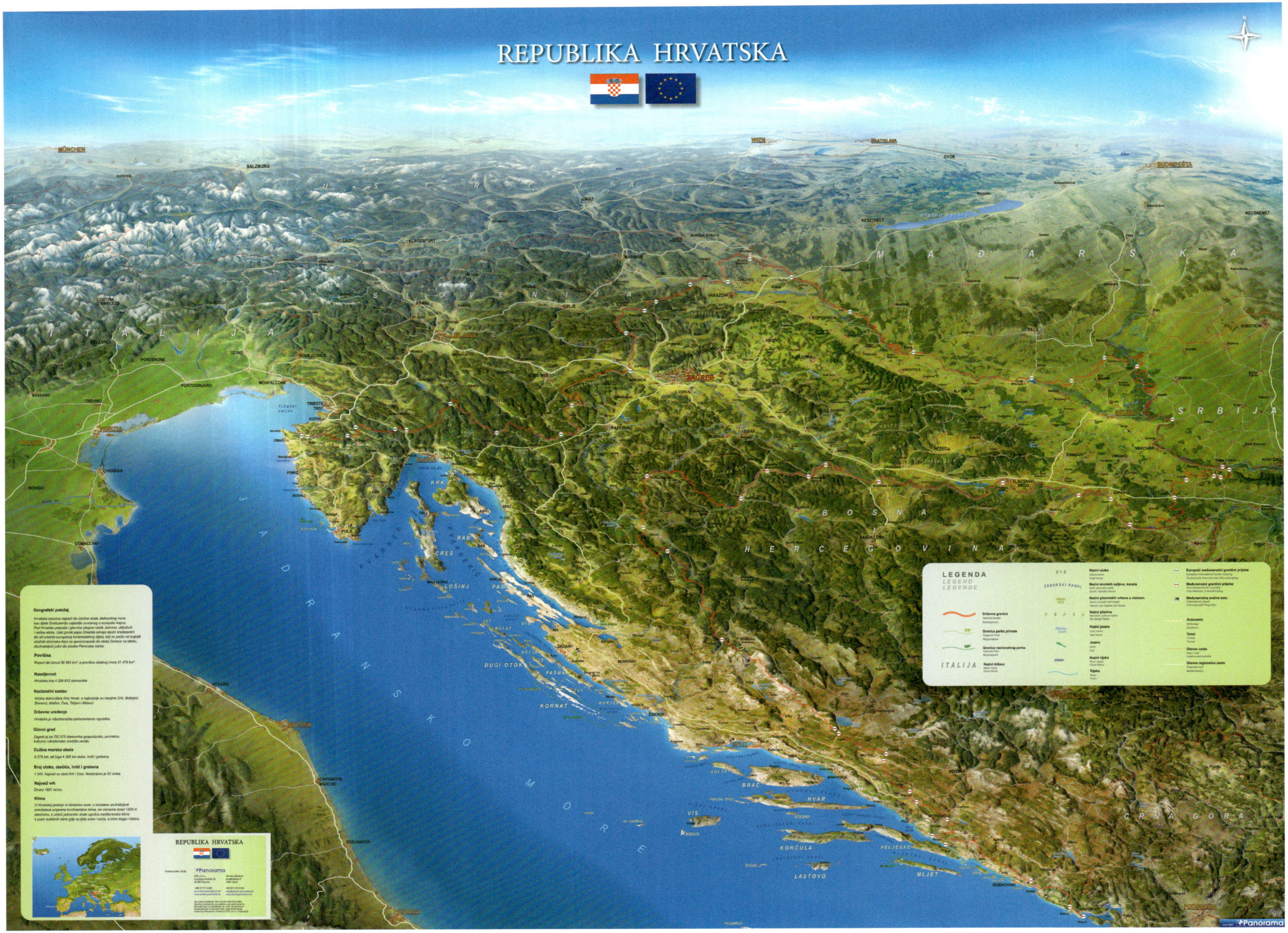

REPUBLIKA HRVATSKA
MÜNCHEN
SALZBURG
WIEN
BRATISLAVA
GYŐR
BUDIMPEŠTA
GRAZ
KESZTHELY
KECSKEMÉT
M A Đ A R S K A
S L O V E N I J A
I T A L I J A
UDINE
PORDENONE
PORTOGRUARO
MONFALCONE
TRIESTE
TREVISO
BASSANO
VENEZIA
CHIOGGIA
ROVIGO
COMACCHIO
ZAGREB
SUBOTICA
S R B I J A
NOVI SAD
B O S N A
H E R C E G O V I N A
C R N A   G O R A
J A D R A N S K O   M O R E
KRK
CRES
LOŠINJ
RAB
PAG
DUGI OTOK
KORNAT
ZADAR
BRAČ
HVAR
VIS
KORČULA
LASTOVO
MLJET
PELJEŠAC
DUBROVNIK
PESARO
LEGENDA
LEGEND
LEGENDE
ITALIJA
REPUBLIKA HRVATSKA
Panorama

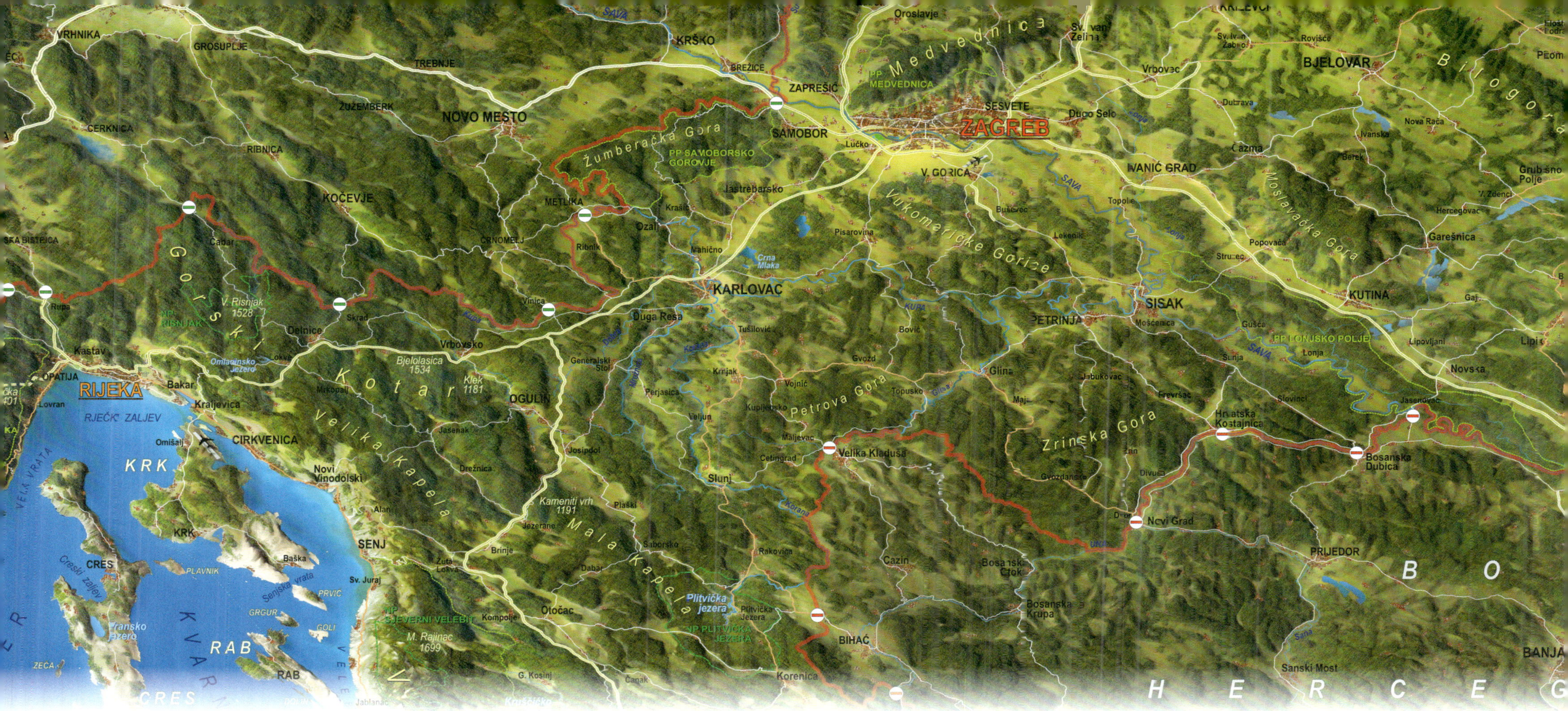

⟪ Panoramic Map of Croatia

Panorama

This panoramic map of Croatia displays the diversity and beauty of the landscape with a unique visual aesthetic. It is content-rich, elaborate, and vividly colorful, yet simple to use. It is not a map in the classic cartographic sense, nor subject to cartographic rules. I drew inspiration from Heinrich Berann, who was the highest authority in the field of panoramic cartography. I analyzed his methods, such as how to "cheat" to get the right forms for the hills and other landscape features. I began with a standard plan-view map, satellite images, and some aerial imagery of the mountains. I wanted to achieve a combination of cartographic accuracy and a panoramic view—something that, as it seems to me, is part of the future of cartography.

All of my work is handmade, with each region of area being manually drawn; I made my first panoramas with a brush. Drawing gives me the freedom to determine which parts of the scene are more or less prominent, and has also been valuable in teaching me about how humans perceive perspective. Despite having this freedom, I still try to minimize its influence, ensuring that the drawn area resembles the real world as much as possible.

Panoramic views are particularly effective at helping people understand space, and viewers find them to be a natural way of experiencing a place. The purpose of my maps is to allow people to look at the land in its natural context.

> *...when designing print maps for outdoor use, you have to consider practical issues like the expectations and skills of the map user.*

The Milford Track »

Roger **Smith**

A field map has to function as such. It helps if it also looks good, but when designing print maps for outdoor use, you have to consider practical issues like the expectations and skills of the map user. This is a print map designed for outdoor use. It is pitched at those who plan to walk the Milford Track, and its purpose is to add value to the visitor experience.

The Milford Track is a 3–4 day wilderness walk in New Zealand. If walked during the recommended summer season it could be described as "soft adventure," a significant commitment and physical challenge for most people, but not requiring any special back-country bushcraft or navigational skills. So Milford Track walkers need not be hard-core hikers. The track is obvious and well sign-posted, and there is shelter to be found along the way.

To achieve its purpose this map had to perform three distinct functions. First up, to serve as a planning aid and pre-walk appetiser, a promise of the good things to come. The bold, cheerful and welcoming colours were chosen to excite and build a sense of anticipation.

Second, the map is needed on the walk itself—used to monitor progress, identify features encountered along the way, and to help users better relate to the surrounding wilderness. I chose here to create a realistic-looking, pseudo-natural base map, which could be intuitively understood and easily interpreted. The track itself needed to be clearly and emphatically represented, so I drew it as a confidence-boosting "yellow brick road": one that, if followed, will take people precisely where they want to go.

Lastly, I wanted to create a map that people would choose to keep afterwards as a souvenir, a tangible reminder of the experience and time spent on the track. So the map had to have character and it needed to be durable. *The Milford Track* is printed on environment-friendly stone paper that is waterproof, insect-proof, and tear-resistant. It is literally printed on rocks.

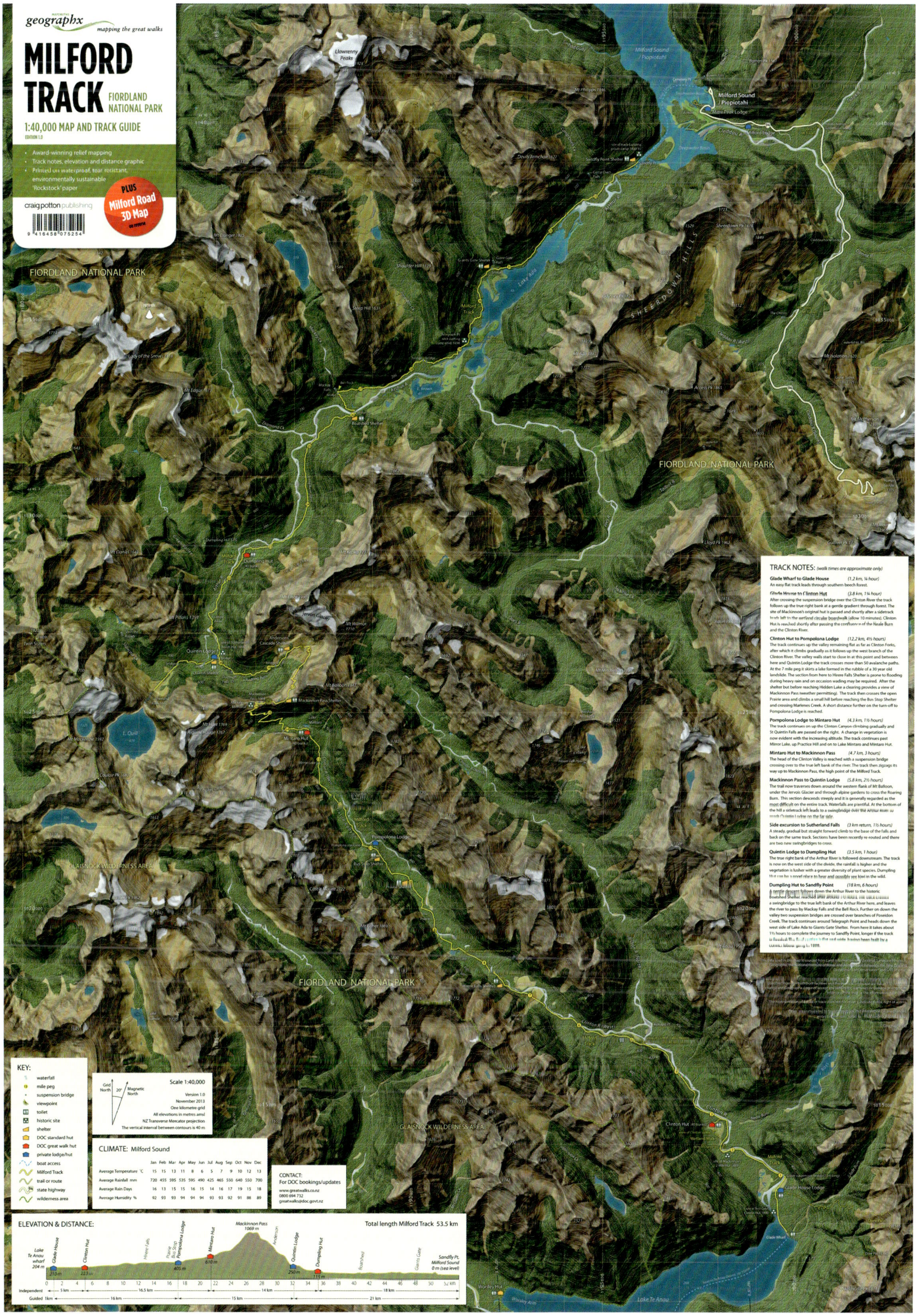

geographx
mapping the great walks

MILFORD TRACK

FIORDLAND NATIONAL PARK

1:40,000 MAP AND TRACK GUIDE

EDITION 1.0

- Award-winning relief mapping
- Track notes, elevation and distance graphic
- Printed on waterproof, tear resistant, environmentally sustainable 'Rockstock' paper

PLUS Milford Road 3D Map on reverse

craig potton publishing

9 416458 075254

TRACK NOTES: *(walk times are approximate only)*

Glade Wharf to Glade House *(1.2 km, ¼ hour)*

An easy flat track leads through southern beech forest.

Glade House to Clinton Hut *(3.8 km, 1¼ hour)*

After crossing the suspension bridge over the Clinton River the track follows up the true right bank at a gentle gradient through forest. The site of Mackinnon's original hut is passed and shortly after a sidetrack leads left to the wetland circular boardwalk (allow 10 minutes). Clinton Hut is reached shortly after passing the confluence of the Neale Burn and the Clinton River.

Clinton Hut to Pompolona Lodge *(12.2 km, 4½ hours)*

The track continues up the valley remaining flat as far as Clinton Forks, after which it climbs gradually as it follows up the west branch of the Clinton River. The valley walls start to close in at this point and between here and Quintin Lodge the track crosses more than 50 avalanche paths. At the 7 mile peg it skirts a lake formed in the rubble of a 30 year old landslide. The section from here to Hirere Falls Shelter is prone to flooding during heavy rain and on occasion wading may be required. After the shelter but before reaching Hidden Lake a clearing provides a view of Mackinnon Pass (weather permitting). The track then crosses the open Prairie area and climbs a small hill before reaching the Bus Stop Shelter and crossing Marlenes Creek. A short distance further on the turn-off to Pompolona Lodge is reached.

Pompolona Lodge to Mintaro Hut *(4.3 km, 1½ hours)*

The track continues on up the Clinton Canyon climbing gradually and St Quintin Falls are passed on the right. A change in vegetation is now evident with the increasing altitude. The track continues past Mirror Lake, up Practice Hill and on to Lake Mintaro and Mintaro Hut.

Mintaro Hut to Mackinnon Pass *(4.7 km, 3 hours)*

The head of the Clinton Valley is reached with a suspension bridge crossing over to the true left bank of the river. The track then zigzags its way up to Mackinnon Pass, the high point of the Milford Track.

Mackinnon Pass to Quintin Lodge *(5.8 km, 2½ hours)*

The trail now traverses down around the western flank of Mt Balloon, under the Jervois Glacier and through alpine gardens to cross the Roaring Burn. This section descends steeply and it is generally regarded as the most difficult on the entire track. Waterfalls are plentiful. At the bottom of the hill a sidetrack left leads to a swingbridge over the Arthur River to reach Quintin Lodge on the far side.

Side excursion to Sutherland Falls *(3 km return, 1½ hours)*

A steady, gradual but straight forward climb to the base of the falls and back on the same track. Sections have been recently re-routed and there are two new swingbridges to cross.

Quintin Lodge to Dumpling Hut *(3.5 km, 1 hour)*

The true right bank of the Arthur River is followed downstream. The track is now on the west side of the divide, the rainfall is higher and the vegetation is lusher with a greater diversity of plant species. Dumpling Hut can be a good place to hear and possibly see kiwi in the wild.

Dumpling Hut to Sandfly Point *(18 km, 6 hours)*

A gentle descent follows down the Arthur River to the historic Boatshed Shelter, reached after around 1½ hours. The track crosses a swingbridge to the true left bank of the Arthur River here, and leaves the river to pass by Mackay Falls and the Bell Rock. Further on down the valley two suspension bridges are crossed over branches of Poseidon Creek. The track continues around Telegraph Point and heads down the west side of Lake Ada to Giants Gate Shelter. From here it takes about 1½ hours to complete the journey to Sandfly Point, longer if the track is flooded. The final section is flat and wide, having been built by a convict labour gang in [illegible].

KEY:

- waterfall
- mile peg
- suspension bridge
- viewpoint
- toilet
- historic site
- shelter
- DOC standard hut
- DOC great walk hut
- private lodge/hut
- boat access
- Milford Track
- trail or route
- state highway
- wilderness area

Grid North 20° Magnetic North

Scale 1:40,000

Version 1.0
November 2013
One kilometre grid
All elevations in metres amsl
NZ Transverse Mercator projection
The vertical interval between contours is 40 m

CLIMATE: Milford Sound

	Jan	Feb	Mar	Apr	May	Jun	Jul	Aug	Sep	Oct	Nov	Dec
Average Temperature °C	15	15	13	11	8	6	5	7	9	10	12	13
Average Rainfall mm	720	455	595	535	595	490	425	465	550	640	550	700
Average Rain Days	16	13	15	15	16	15	14	16	17	19	15	18
Average Humidity %	92	93	93	94	94	94	93	93	92	91	88	89

CONTACT:
For DOC bookings/updates
www.greatwalks.co.nz
0800 694 732
greatwalks@doc.govt.nz

ELEVATION & DISTANCE:

Total length Milford Track 53.5 km

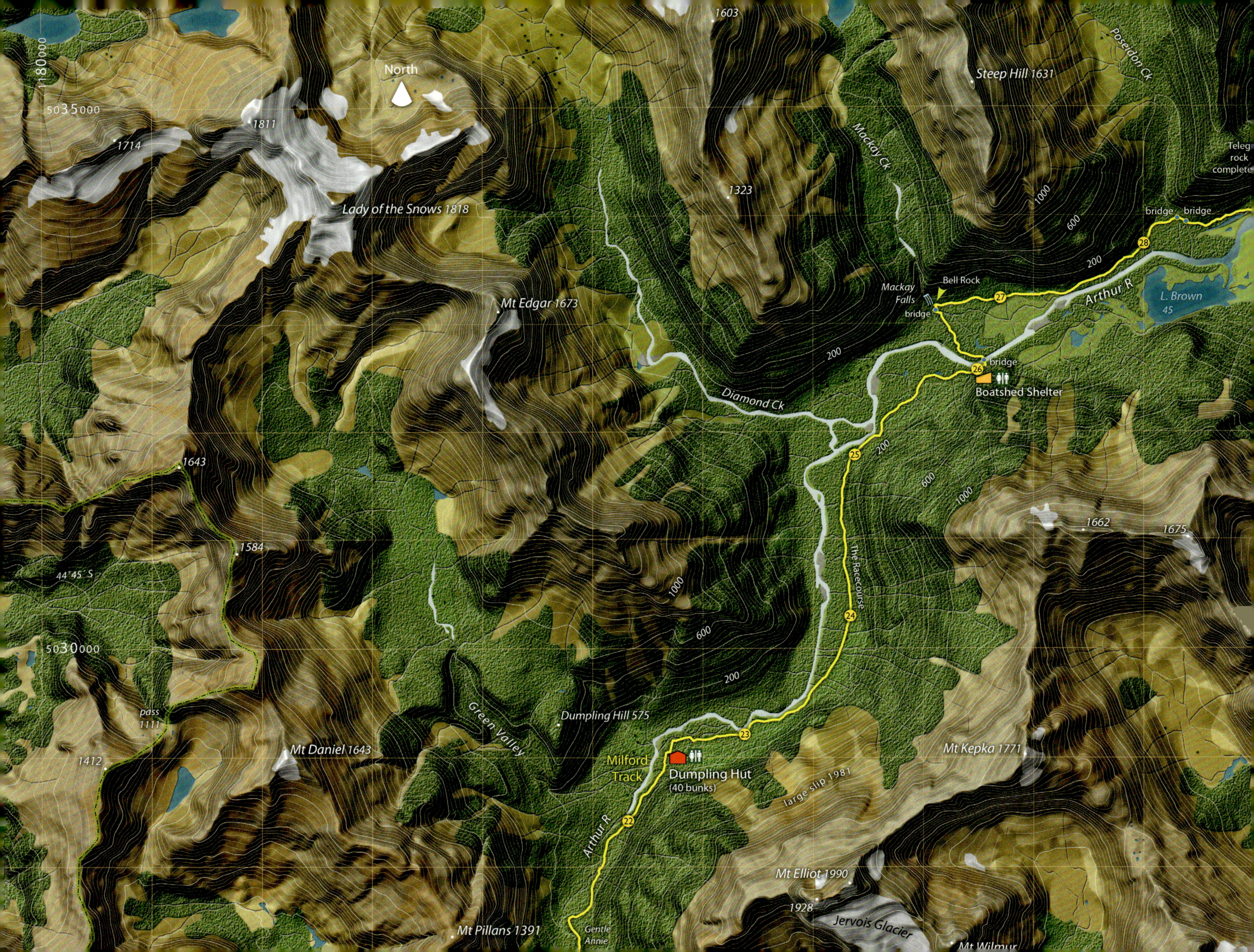

North
1603
Steep Hill 1631
Poseidon Ck
5035000
1811
1714
Lady of the Snows 1818
Mackay Ck
1323
bridge
bridge
1000
600
200
28
Mackay Falls
Bell Rock
bridge
27
Arthur R
L. Brown
45
Mt Edgar 1673
200
Diamond Ck
bridge
26
Boatshed Shelter
200
25
600
1000
1643
1662
1675
1584
44°45'S
The Racecourse
1000
24
600
200
5030000
pass
1111
Dumpling Hill 575
Green Valley
Mt Daniel 1643
23
Milford Track
Dumpling Hut
(40 bunks)
Mt Kepka 1771
large slip 1981
1412
Arthur R
22
Mt Elliot 1990
1928
Jervois Glacier
Mt Pillans 1391
Gentle Annie

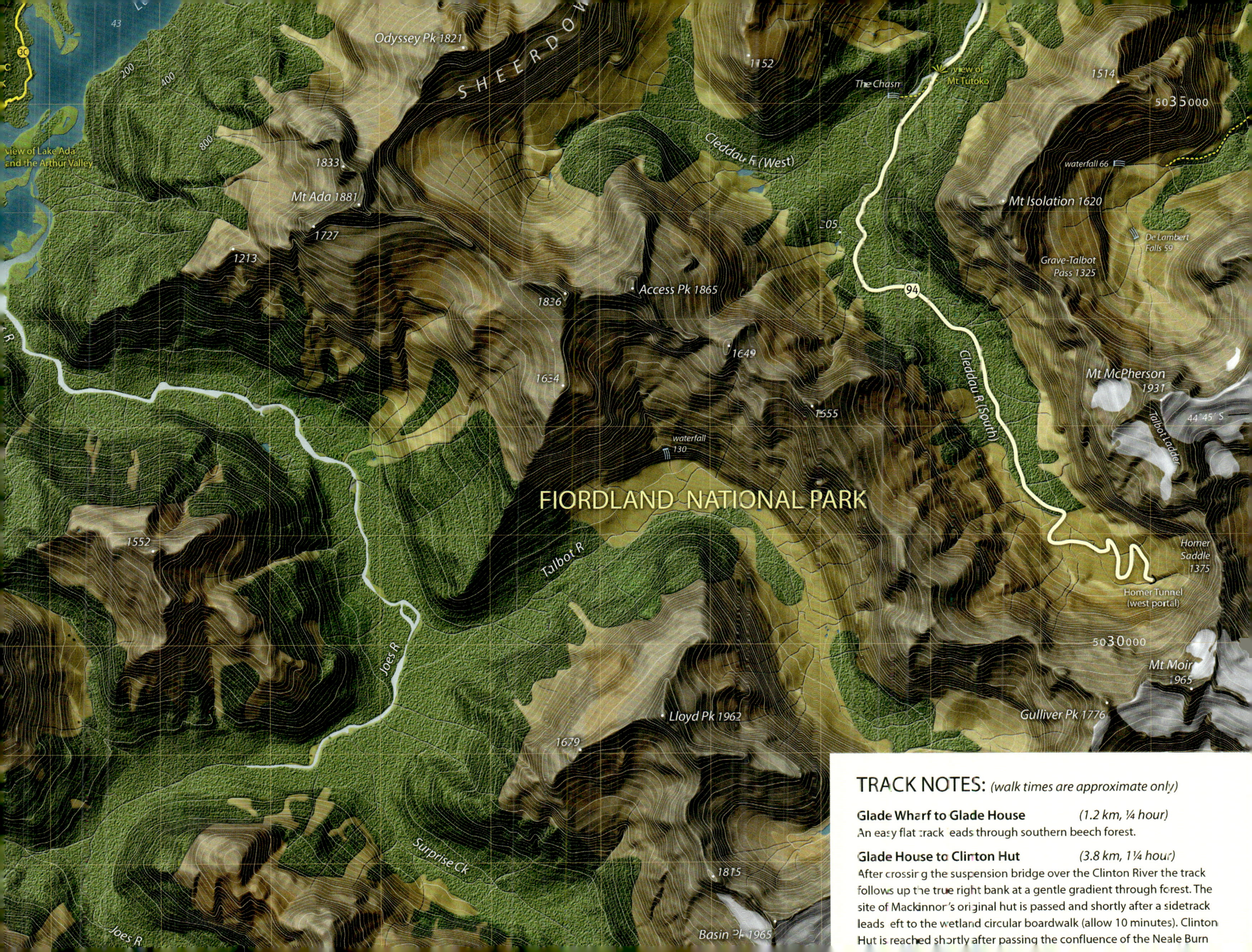

TRACK NOTES: *(walk times are approximate only)*

Glade Wharf to Glade House *(1.2 km, ¼ hour)*

An easy flat track leads through southern beech forest.

Glade House to Clinton Hut *(3.8 km, 1¼ hour)*

After crossing the suspension bridge over the Clinton River the track follows up the true right bank at a gentle gradient through forest. The site of Mackinnon's original hut is passed and shortly after a sidetrack leads left to the wetland circular boardwalk (allow 10 minutes). Clinton Hut is reached shortly after passing the confluence of the Neale Burn

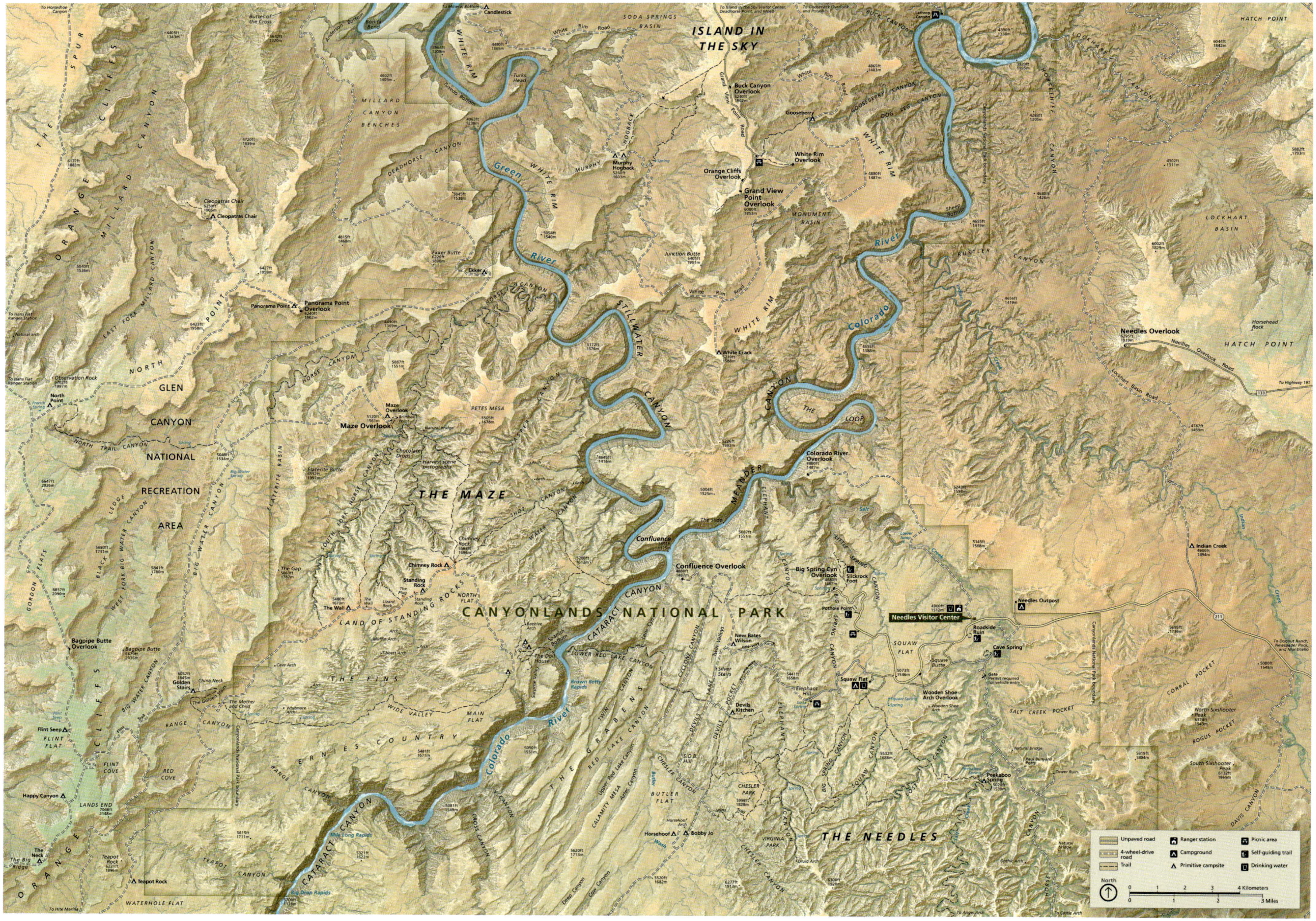

ISLAND IN THE SKY
SODA SPRINGS BASIN
HATCH POINT
THE SPUR
ORANGE CLIFFS
MILLARD CANYON
MILLARD CANYON BENCHES
DEADHORSE CANYON
Green River
WHITE RIM
MURPHY HOGBACK
Murphy Hogback
Buck Canyon Overlook
Gooseberry
White Rim Overlook
Orange Cliffs Overlook
Grand View Point Overlook
MONUMENT BASIN
Colorado River
LOCKHART CANYON
LOCKHART BASIN
KERSHAW CANYON
Candlestick
Cleopatras Chair
Ekker Butte
Junction Butte
Panorama Point
Panorama Point Overlook
NORTH POINT
North Point
STILLWATER CANYON
White Crack
Needles Overlook
Horsehead Rock
GLEN CANYON NATIONAL RECREATION AREA
Maze Overlook
PETES MESA
THE MAZE
THE LOOP
Colorado River Overlook
MEANDER CANYON
Confluence
Confluence Overlook
Chimney Rock
Standing Rock
The Wall
The Gap
LAND OF STANDING ROCKS
CANYONLANDS NATIONAL PARK
Big Spring Cyn Overlook
Slickrock Foot
Needles Visitor Center
Needles Outpost
Indian Creek
Pothole Point
Roadside Ruin
SQUAW FLAT
Squaw Flat
Cave Spring
New Bates Wilson
Bagpipe Butte Overlook
Bagpipe Butte
Golden Stairs
THE FINS
WIDE VALLEY
MAIN FLAT
CATARACT CANYON
Brown Betty Rapids
ERNIES COUNTRY
Elephant Hill
Devils Kitchen
Wooden Shoe Arch Overlook
SALT CREEK POCKET
CORRAL POCKET
North Sixshooter Peak
BOGUS POCKET
South Sixshooter Peak
Flint Seep
FLINT FLAT
FLINT COVE
RED COVE
Happy Canyon
LANDS END
THE GRABENS
CHESLER PARK
BUTLER FLAT
Horsehoof Arch
Bobby Jo
Peekaboo
THE NEEDLES
Teapot Rock
TEAPOT CANYON
WATERHOLE FLAT
The Neck
The Big Ridge
Mile Long Rapids
Big Drop Rapids
VIRGINIA PARK
DAVIS CANYON
Unpaved road
4-wheel-drive road
Trail
Ranger station
Campground
Primitive campsite
Picnic area
Self-guiding trail
Drinking water
North
0 1 2 3 4 Kilometers
0 1 2 3 Miles

« The Heart of Canyonlands National Park

Tom **Patterson**

Recreating the rock textures of hand-drawn Swiss topographic sheets in an automated way has been an elusive goal of cartographers. I made this map as an experiment to demonstrate the rock depiction capabilities of Terrain Texture Shader, a new application that detects, amplifies, and renders slight variations in digital elevation models as raster textures resembling rocks. The texture shading technique works best on landscapes with horizontal strata, such as that of Canyonlands National Park, Utah.

Another factor for selecting Canyonlands is my admiration for Bradford Washburn's 1978 map *The Heart of the Grand Canyon*, co-produced with *National Geographic* and with contributions from SwissTopo. This classic map masterfully applies Swiss-style rock hachuring, originally developed for mapping alpine peaks, to the "inverted" canyon topography of the Colorado River. Canyonlands National Park, upriver from the Grand Canyon, was virgin territory for making a similar map.

One challenge was showing the intricate terrain of Canyonlands while not having it overwhelm the map with visual noise. Blending shaded relief with texture shading (rendered at the default settings) yielded the desired balance. The inclusion of texture shading also enhances the three-dimensional appearance of shaded relief. This technique avoids the potential problem of relief inversion, such as when canyons appear as ridges, and vice versa. The rock textures that I rendered separately appear only on the steepest slopes.

The map colors derive from aerial photography merged with the relief renderings described above. Color adjustments to the aerial photographs included changing the dark green woodland found at upper elevations to lime green. This tweak conforms to the "higher is lighter" elevation color scheme to better distinguish the highlands from lowlands. I also show the muddy waters of the Colorado and Green rivers as blue—my favorite color.

Washburn's *Heart of the Grand Canyon* took seven years to produce and involved arduous field surveys by his family and volunteers. By comparison, my *Heart of Canyonlands National Park* took 32 hours to complete, thanks entirely to the efficient mapping software and data that I used.

One challenge was showing the intricate terrain of Canyonlands while not having it overwhelm the map with visual noise.

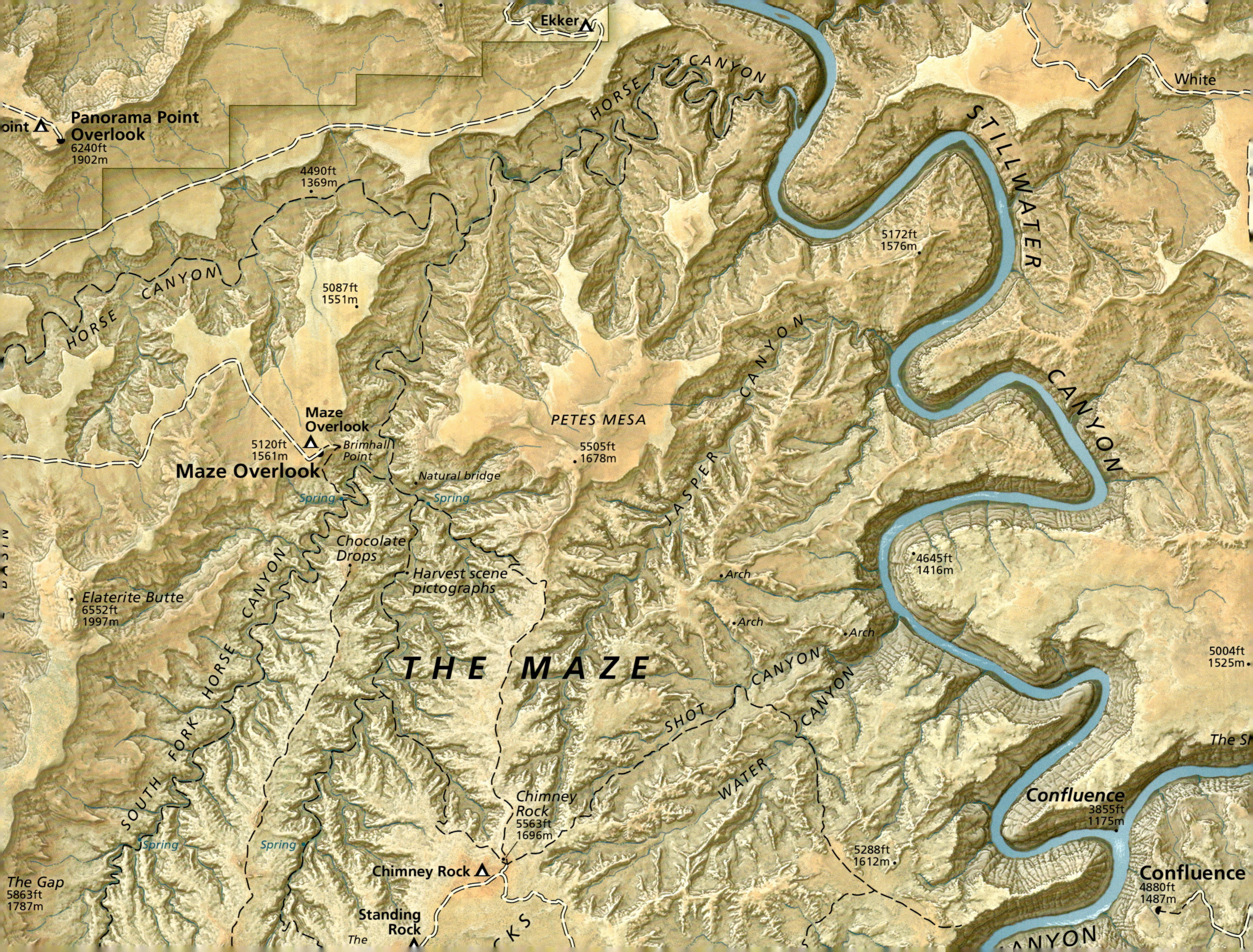
Ekker
Panorama Point Overlook
6240ft
1902m
White
HORSE CANYON
STILLWATER CANYON
4490ft
1369m
5172ft
1576m
HORSE CANYON
5087ft
1551m
JASPER CANYON
PETES MESA
Maze Overlook
5120ft
1561m
Brimhall Point
Maze Overlook
5505ft
1678m
Natural bridge
Spring
Spring
Chocolate Drops
Harvest scene pictographs
4645ft
1416m
Arch
Arch
Arch
Elaterite Butte
6552ft
1997m
SOUTH FORK HORSE CANYON
THE MAZE
SHOT CANYON
WATER CANYON
5004ft
1525m
The S
Confluence
3855ft
1175m
Chimney Rock
5563ft
1696m
Spring
Spring
5288ft
1612m
Chimney Rock
Confluence
4880ft
1487m
The Gap
5863ft
1787m
Standing Rock
The
CANYON

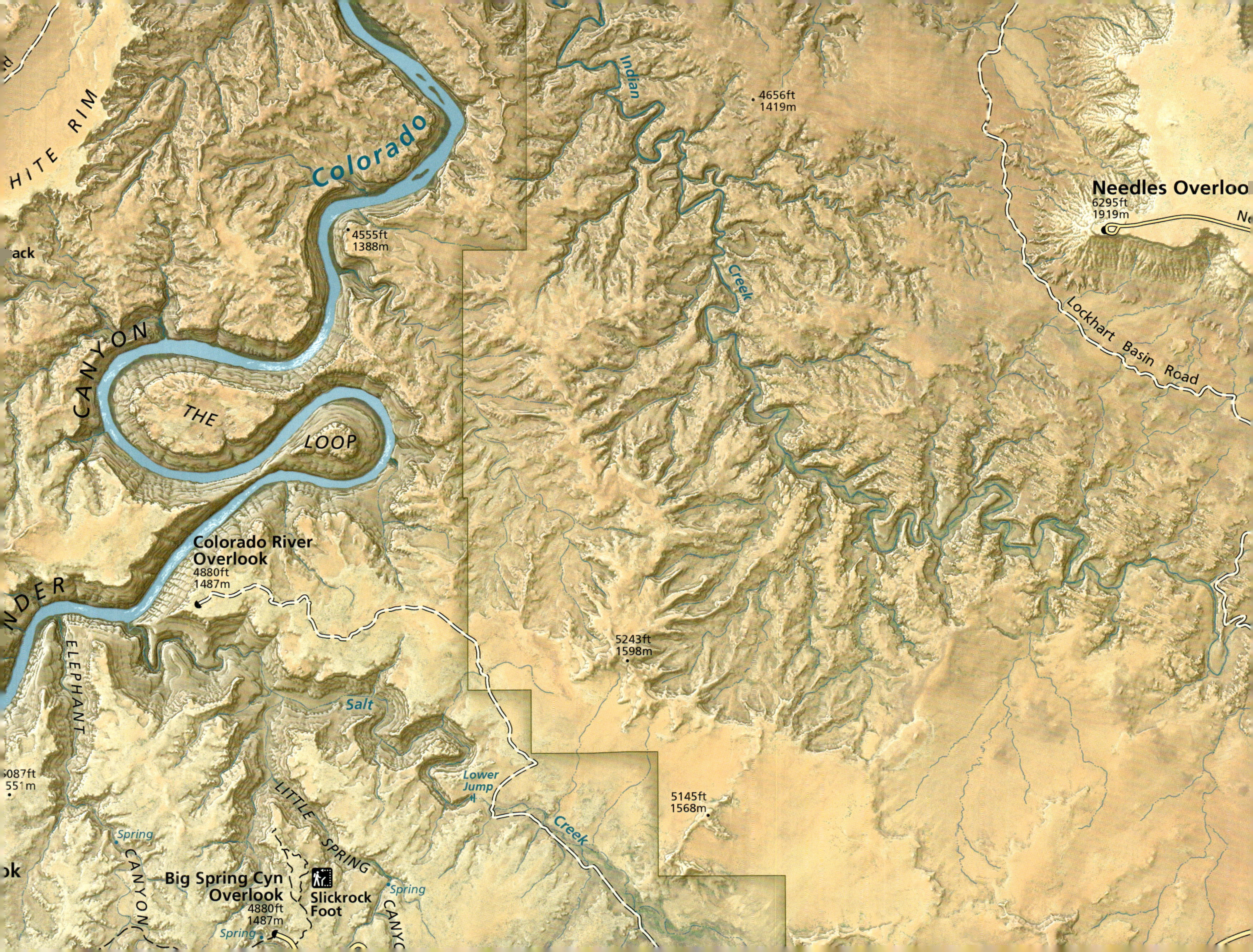

HITE RIM
Colorado
Indian
Creek
4656ft
1419m
4555ft
1388m
CANYON
THE
LOOP
Needles Overloo
6295ft
1919m
Lockhart Basin Road
Colorado River
Overlook
4880ft
1487m
NDER
ELEPHANT
Salt
5243ft
1598m
Lower
Jump
5145ft
1568m
Creek
LITTLE
SPRING
Spring
CANYON
Big Spring Cyn
Overlook
4880ft
1487m
Slickrock
Foot
Spring
Spring

» *What started out geometric and widely spaced became smooth and intricate as I tried to replicate the appearance of flowing land.*

Mt. Hood »

Kelly ***Abplanalp***

I began with pen, paper, and an ambition to replicate a standard hachuring technique, but my initial sketches and the final product differ greatly. What started out geometric and widely spaced became smooth and intricate as I tried to replicate the appearance of flowing land. Ultimately these attempts gave rise to the distinctive pattern that serves as a base for my terrain maps, characterized by elongated triangles and V shapes moving with the land. For example, a common arrangement is to place the short faces of the triangles against each other along a ridge top while their tails run down into the valley. The end result is intriguing to look at, but it is impossible to gain a sense of depth from a uniformly black and white image. I shaded the map using Photoshop to create the illusion of elevation.

I aim to depict Mt. Hood in a way that will instill a sense of beauty and curiosity in the reader. The pattern of the land is the focus of my work, but in addition to the hillshade and ink pattern, I included a few important features: Timberline Trail, glacier labels, and hand-drawn symbols designed with cohesion to the background style in mind. Consistency of design is also present in the border of the locator map, which mimics patterns seen in the landscape.

Mountains and their connections to people are powerful, and I hope that this map of Mt. Hood will help readers understand the land and explore the mountain while appreciating the intricacies of its terrain.

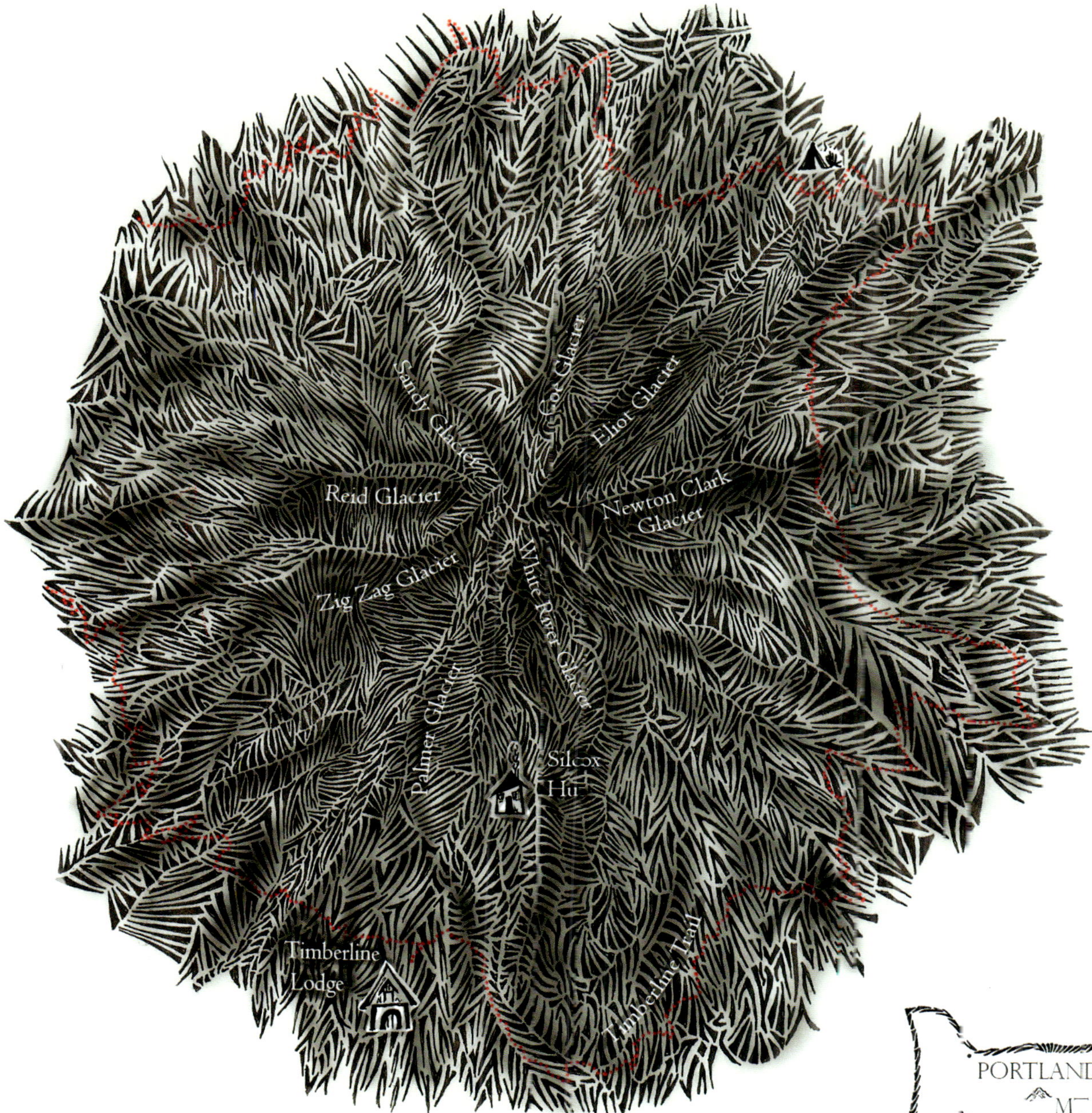
Sandy Glacier
Coe Glacier
Eliot Glacier
Reid Glacier
Newton Clark
Glacier
Zig Zag Glacier
White River Glacier
Palmer Glacier
Silcox
Hu
Timberline
Lodge
Timberline Trail

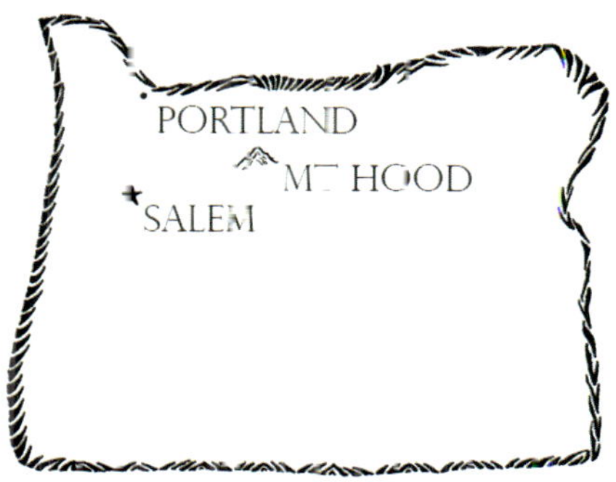
PORTLAND
MT HOOD
SALEM

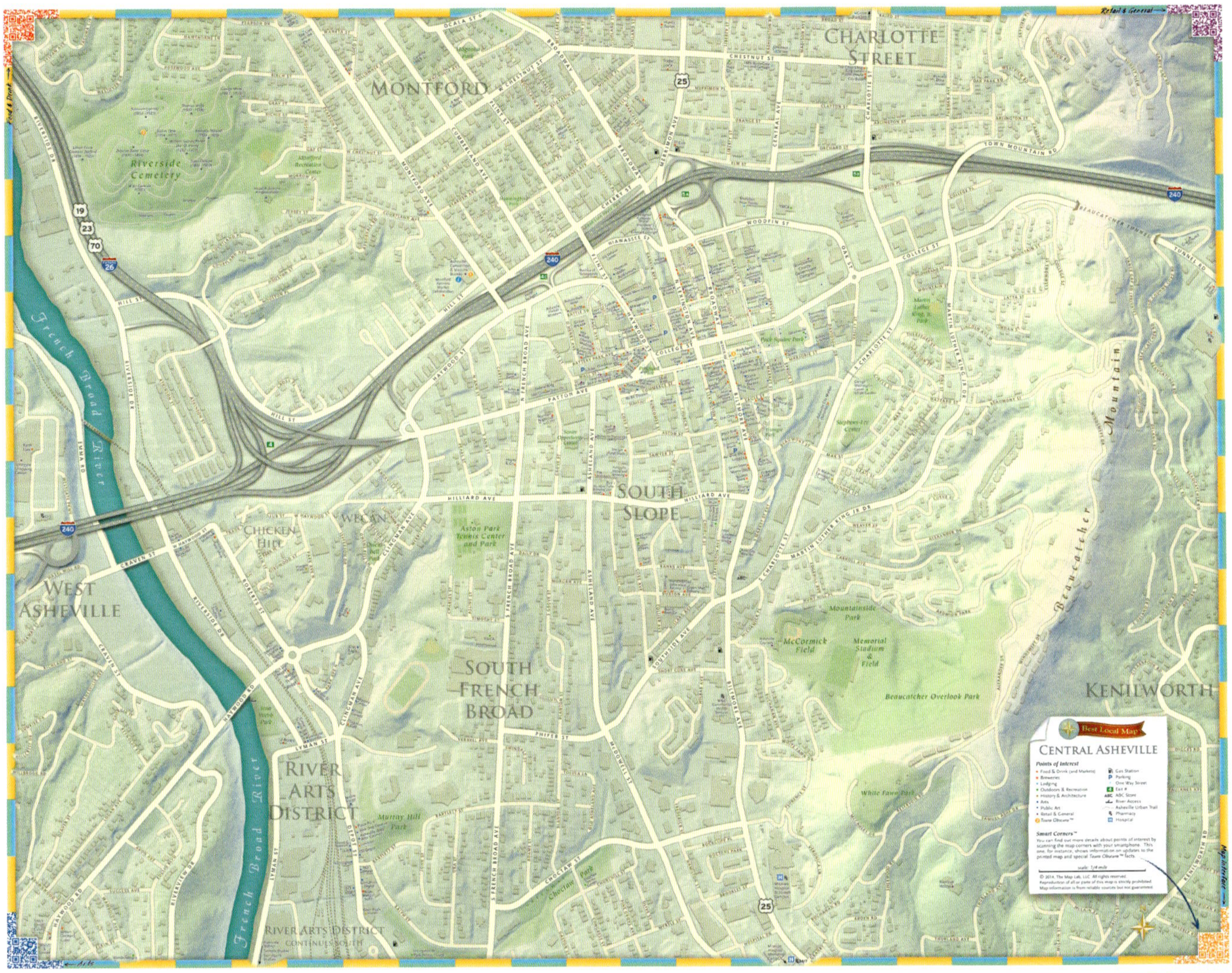

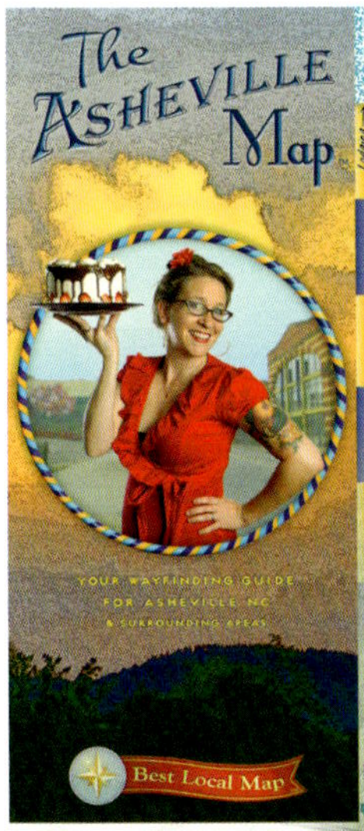

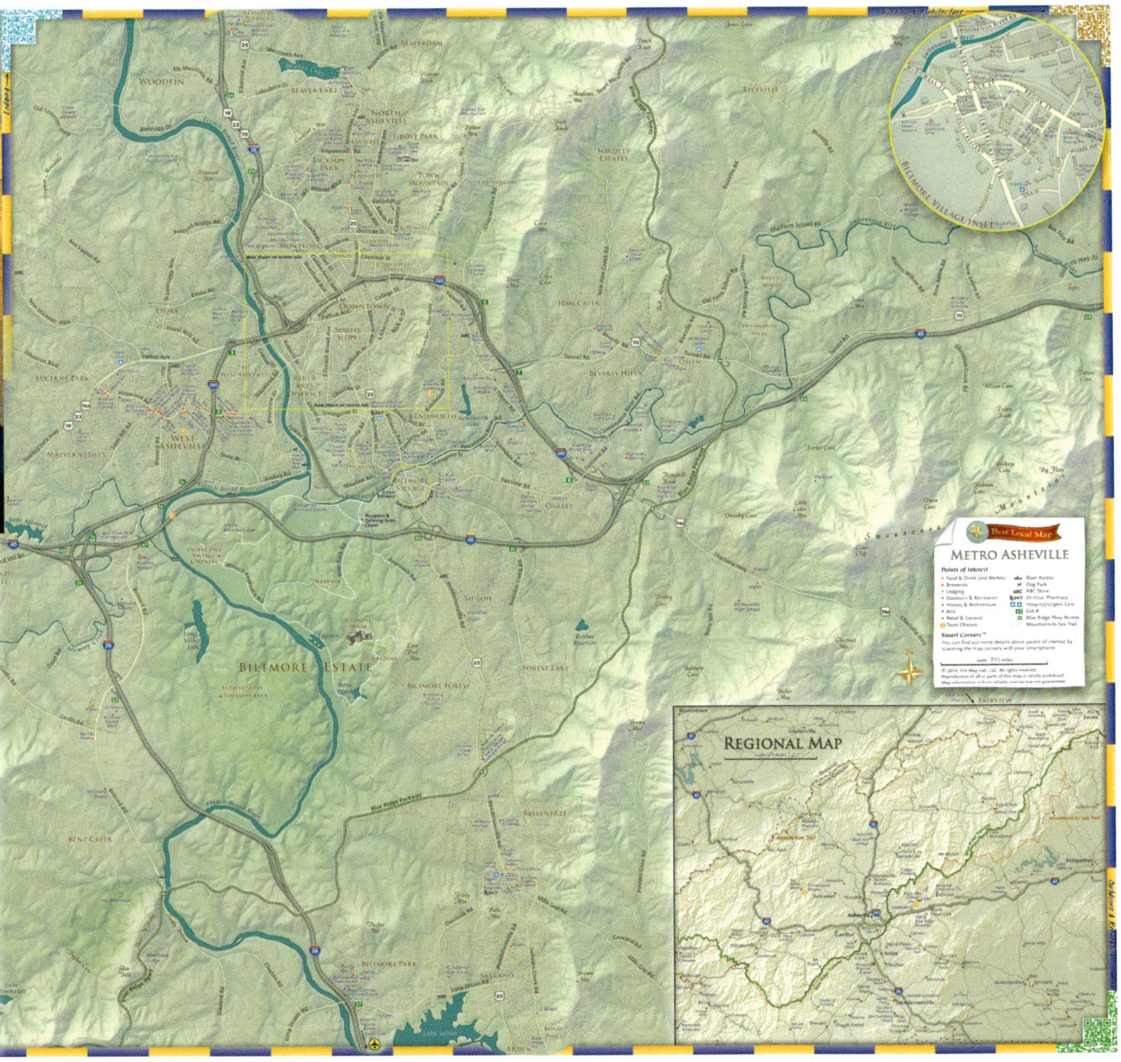

The Asheville Map™

· TOPOGRAPHIC DETAIL · TRUE-TO-SCALE
· SMART CORNERS™ · NO ADVERTISING · BREWERIES !

Best Local Map

More than a map

Best Local Map celebrates the unique qualities, attractions, and local flavor of cities that are small in size and big at heart. This map is an invitation to explore the area using Smart Corners™ as an additional guide. In the spirit of sharing Asheville's bounty with everyone, each sale of The Asheville Map gives back to this community. ***www.bestlocalmap.com***

MANNA FOODBANK

More than a meal

MANNA FoodBank is the designated beneficiary of The Asheville Map. MANNA's mission is to involve, educate, and unite people in the work of ending hunger in Western North Carolina.

A portion of funds from each map sold will enable MANNA to provide food for more than one meal.

Together we can solve hunger.

Smart Corners™

Points of interest on the map are color-coded by category. For detailed listings of local offerings and easy access to websites in each category, scan the matching corner of the map with a QR reader app. For **updates** and map information, scan the lower right corner of the Central Asheville map, or go to: www.bestlocalmap.com/Asheville/Updates

Toura Obscura™

The Toura Obscura™ is a portal into the fabric of Asheville, a jumping-off point for self-guided tours, and a work in progress. A series of numbered markers on the map pinpoint places to see, curiosities, or stories to consider. Each marker has a corresponding narrative available through the lower right Smart Corner™ on the Central Asheville map or: www.bestlocalmap.com/Asheville/TouraObscura

Produced and published by The Map Lab, LLC

ISBN 978-0-9910277-0-5

« The Asheville Map

Bruce ***Daniel*** *and Nora* ***Daniel*** *— Best Local Map*

Ever since we first visited Asheville, North Carolina, my wife and I wanted a map that reflected the unique fabric of the town. Since that map didn't exist, we decided to create it.

My list of must-haves included topography, lots of our favorite spots, and a level of detail that revealed the richness of neighborhoods, restaurants, breweries, and activities that keep locals proud and visitors returning. Honoring the spirit of this community for us includes giving over 20 percent of wholesale revenue from map sales to MANNA FoodBank, a nonprofit working to end hunger in western North Carolina.

I started this map at an important juncture in my career, and found myself reviewing a compendium of my previous map designs. I came across a design mockup I felt had remained unfinished and wanted to see where it might lead—in particular, exploring grey highways, warm secondary roads, dashed centerlines, nuanced drop shadows, and casings.

Design inspiration came from a number of sources. The base color drew from the surrounding Blue Ridge Mountains, where on every horizon, a bluish, purplish green blends with a hazy atmosphere. The cover imagery, layout, and map borders were inspired by vintage circus posters and the ever-present quirky amusement found on the streets of Asheville. Great inspiration was taken from Charles Booth's 1889 London map of poverty and wealth, *Life and Labour of the People*. I've often found myself engaged in debate about the merits of using many different fonts, and found in Booth's London map support for the use of a large variety of type styles and sizes, especially small type for details. I love the way Booth broke his map edge for elements and labels and found that my doing the same promoted the unexpected feel I was after.

While we clearly wanted to produce a paper product, we also wanted our map and its 500 curated points of interest to link to additional information. This led to me incorporating graphically designed QR codes into the map border, deemed Smart Corners™, effectively extending the map into the digital realm and keeping the information on it current.

> *Great inspiration was taken from Charles Booth's 1889 London map of poverty and wealth,* Life and Labour of the People.

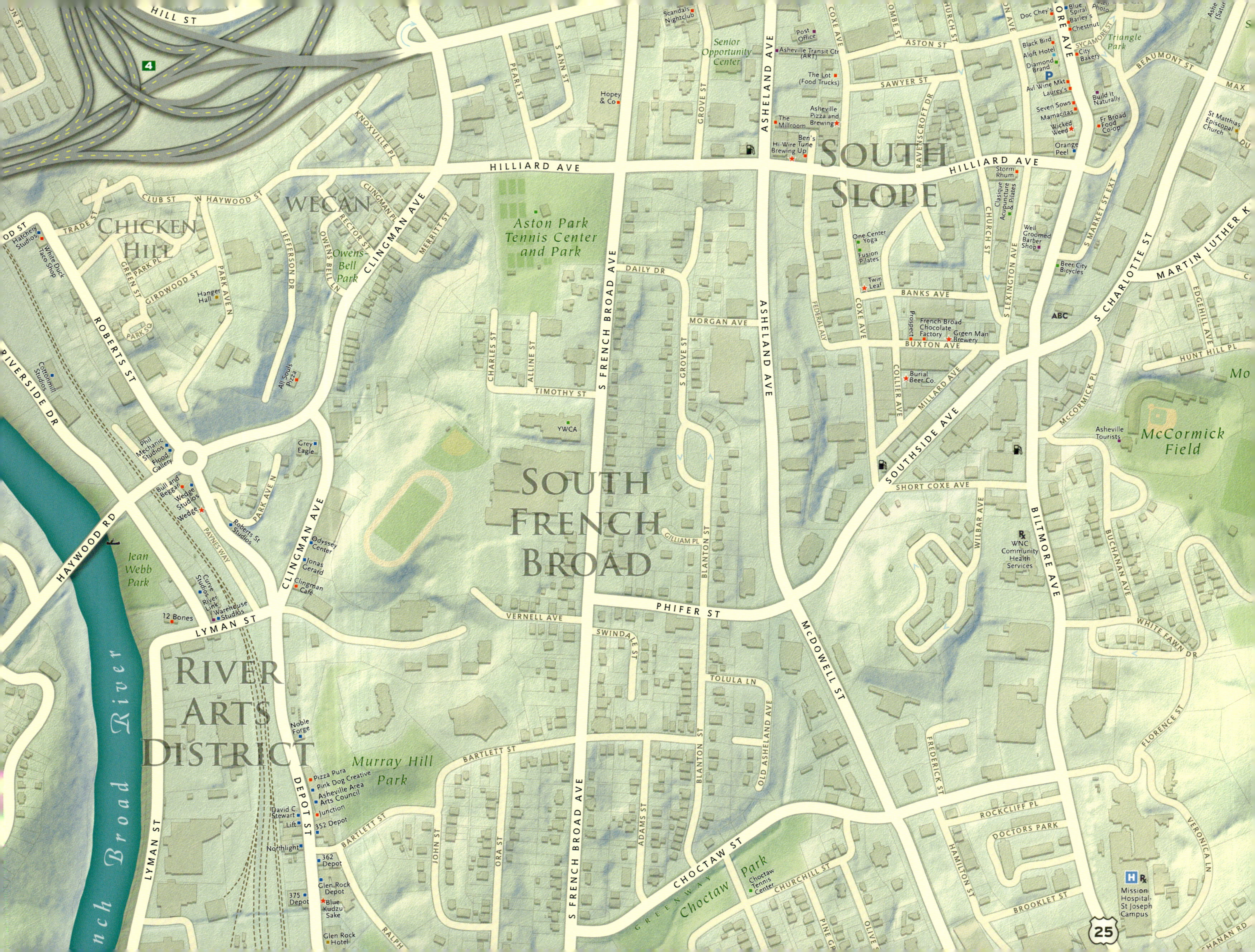
HILL ST
4
KNOXVILLE PL
PEARL ST
S ANN ST
Hopey & Co
GROVE ST
Scandals Nightclub
Senior Opportunity Center
Post Office
Asheville Transit Ctr (ART)
The Lot (Food Trucks)
ASHELAND AVE
COXE AVE
ASTON ST
SAWYER ST
The Millroom
Asheville Pizza and Brewing
Ben's
Hi-Wire Brewing
Tune Up
SOUTH SLOPE
RAVENSCROFT DR
HILLIARD AVE
Doc Chey's
Black Bird
Aloft Hotel
Diamond Brand
Avl Wine Mkt
Laurey's
Seven Sows
Mamacitas
Wicked Weed
Orange Peel
City Bakery
Build It Naturally
Fr Broad Food Co-op
Triangle Park
SYCAMORE ST
BEAUMONT ST
MAX
St Matthias Episcopal Church
CHICKEN HILL
WECAN
CLUB ST
W HAYWOOD ST
TRADE ST
Hatchery Studios
White Duck Taco Shop
PARK PL
GREEN ST
GIRDWOOD ST
Hanger Hall
PARK AVE N
JEFFERSON DR
RECTOR ST
CLINGMAN PL
OWENS BELL LN
Owens-Bell Park
CLINGMAN AVE
MERRITT ST
Aston Park Tennis Center and Park
S FRENCH BROAD AVE
DAILY DR
MORGAN AVE
Storm Rhum
Clasique Acupuncture & Pilates
One Center Yoga
Fusion Pilates
Twin Leaf
BANKS AVE
Well Groomed Barber Shop
Beer City Bicycles
S MARKET ST EXT
S LEXINGTON AVE
CHURCH ST
S CHARLOTTE ST
MARTIN LUTHER K
ABC
FEDERAL ALY
Prospect
French Broad Chocolate Factory
Green Man Brewery
BUXTON AVE
Burial Beer Co.
COLLIER AVE
MILLARD AVE
EDGEHILL AVE
HUNT HILL PL
McCORMICK PL
Asheville Tourists
McCormick Field
ROBERTS ST
RIVERSIDE DR
Cotton Mill Studios
All Souls Pizza
CHARLES ST
ALLINE ST
TIMOTHY ST
YWCA
S GROVE ST
SOUTHSIDE AVE
SHORT COXE AVE
Grey Eagle
Phil Mechanic Studios
Flood Gallery
Bull and Beggar
Wedge Studios
Wedge
Roberts St Studios
PAYNES WAY
PARK AVE N
SOUTH FRENCH BROAD
GILLIAM PL
BLANTON ST
WILBAR AVE
WNC Community Health Services
BILTMORE AVE
BUCHANAN AVE
HAYWOOD RD
Jean Webb Park
Curve Studios
River Link
Warehouse Studios
Odyssey Center
Jonas Gerard
Clingman Cafe
12 Bones
LYMAN ST
VERNELL AVE
PHIFER ST
SWINDALE ST
McDOWELL ST
WHITE FAWN DR
RIVER ARTS DISTRICT
French Broad River
TOLULA LN
OLD ASHELAND AVE
FREDERICK ST
FLORENCE ST
Noble Forge
Murray Hill Park
BARTLETT ST
Pizza Pura
Pink Dog Creative
Asheville Area Arts Council
Junction
David C. Stewart
Lift
352 Depot
DEPOT ST
Northlight
362 Depot
375 Depot
Glen Rock Depot
Blue Kudzu Sake
Glen Rock Hotel
RALPH
JOHN ST
ORA ST
ADAMS ST
CHOCTAW ST
Choctaw Park
Choctaw Tennis Center
GREENWAY
CHURCHILL ST
HAMILTON ST
ROCKCLIFF PL
DOCTORS PARK
BROOKLET ST
Mission Hospital-St Joseph Campus
VERONICA LN
25

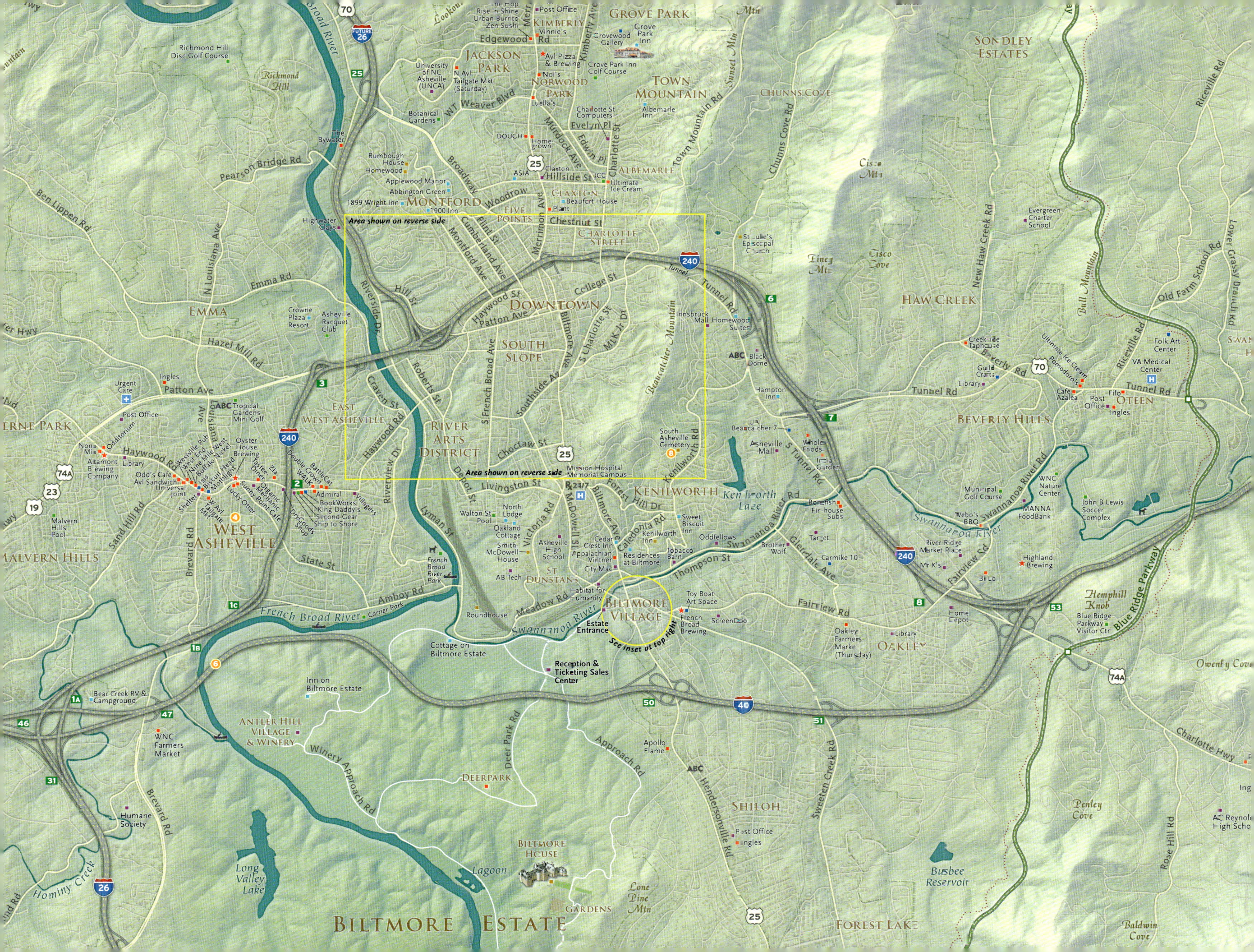

Grove Park
Kimberly
Jackson Park
Norwood Park
Town Mountain
Sondley Estates
Chunns Cove
Albemarle
Claxton
Montford
Five Points
Charlotte Street
Area shown on reverse side
Downtown
South Slope
River Arts District
East West Asheville
Emma
Haw Creek
Beverly Hills
Oteen
Kenilworth
Kenilworth Lake
West Asheville
Malvern Hills
Biltmore Village
See inset at top right
St Dunstans
Oakley
Shiloh
Forest Lake
Biltmore Estate
Biltmore House
Gardens
Lagoon
Deerpark
Antler Hill Village & Winery
Inn on Biltmore Estate
Cottage on Biltmore Estate
Reception & Ticketing Sales Center
Estate Entrance
Long Valley Lake
Lone Pine Mtn
Busbee Reservoir
Penley Cove
Baldwin Cove
Owenby Cove
Hemphill Knob
Cisco Mtn
Cisco Cove
Piney Mtn
Richmond Hill
Richmond Hill Disc Golf Course
French Broad River
Swannanoa River
Hominy Creek
Blue Ridge Parkway
Broad River
Pearson Bridge Rd
Ben Lippen Rd
Emma Rd
Hazel Mill Rd
Patton Ave
Haywood Rd
Riverside Dr
Hill St
Craven St
Roberts St
Haywood St
Patton Ave
College St
Chestnut St
Merrimon Ave
Broadway
Montford Ave
Cumberland Ave
Elliot St
Woodrow
Hillside St
Charlotte St
Town Mountain Rd
Chunns Cove Rd
Tunnel Rd
S French Broad Ave
Biltmore Ave
Southside Ave
S Charlotte St
MLK Jr Dr
Beaucatcher Mountain
Choctaw St
Livingston St
Depot St
Lyman St
McDowell St
Victoria Rd
Meadow Rd
Amboy Rd
Brevard Rd
Sand Hill Rd
State St
Louisiana Ave
N Louisiana Ave
Riverview Dr
Haywood Rd
Kenilworth Rd
Forest Hill Dr
Caledonia Rd
Thompson St
Swannanoa River Rd
Glendale Ave
Fairview Rd
Fairview St
Beverly Rd
Riceville Rd
New Haw Creek Rd
Old Farm School Rd
Lower Grassy Branch Rd
Bull Mountain
Charlotte Hwy
Hendersonville Rd
Sweeten Creek Rd
Rose Hill Rd
Approach Rd
Winery Approach Rd
Deer Park Rd
Brevard Rd
W.T. Weaver Blvd
Edgewood Rd
Evelyn Pl
Murdock Ave
Kimberly Ave
Sunset Mtn
Mission Hospital Memorial Campus
Asheville Mall
Innsbruck Mall
VA Medical Center
WNC Farmers Market
WNC Nature Center
Municipal Golf Course
John B Lewis Soccer Complex
Highland Brewing
French Broad Brewing
Roundhouse
French Broad River Park
Carrier Park
Bear Creek RV & Campground
Humane Society
Apollo Flame
University of NC Asheville (UNCA)
Botanical Gardens
Crowne Plaza Resort
Asheville Racquet Club
Tropical Gardens Mini Golf
Post Office
Library
Ingles
Target
Home Depot
Toy Boat Art Space
ScreenDoor
Oakley Farmers Market (Thursday)
Blue Ridge Parkway Visitor Ctr
South Asheville Cemetery
Asheville High School
Residences at Biltmore
Sweet Biscuit Inn
Oddfellows
Tobacco Barn
Carmike 10
Brother Wolf
Grove Park Inn
Crovewood Gallery
Grove Park Inn Golf Course
Town Mountain
Charlotte St Computers
Albemarle Inn
Ultimate Ice Cream
Beaufort House
1899 Wright Inn
1900 Inn
Abbington Green
Applewood Manor
Rumbough House
Homewood
Highwater Clays
Bywater
Evergreen Charter School
St Lukes Episcopal Church
Folk Art Center
Hampton Inn
Black Dome
Homewood Suites
MANNA FoodBank
Webo's BBQ
Oakland Cottage
Smith-McDowell House
North Lodge
Walton St Pool
Malvern Hills Pool
Altamont Brewing Company
Habitat for Humanity
AB Tech
Cedar Crest Inn
Kenilworth Inn
Lookout
Post Office
Rise 'n Shine
Urban Burrito
Zen Sushi
Vinnie's
Avl Pizza & Brewing
Noi's
Luella's
N Avl Tailgate Mkt (Saturday)
DOUCH
Homegrown
ASIA
Claxton
Plant
Cafe Azalea
Ultimate Ice Cream
Pomodoros
Filo
Guild Crafts
Creekside Taphouse
Whole Foods
India Garden
Boneshed Firehouse Subs
River Ridge Market Place
Mr K's
3FLo
Urgent Care
Oyster House Brewing
Admiral
BookWorks
King Daddy's
Second Gear
Ship to Shore
Villagers
Odd's Cafe
Universal Joint
Sunny Point Cafe
Lucky Otter
Orbit DVD
BattleCat
Dry Goods Shop
Nona Mia
Odditorium

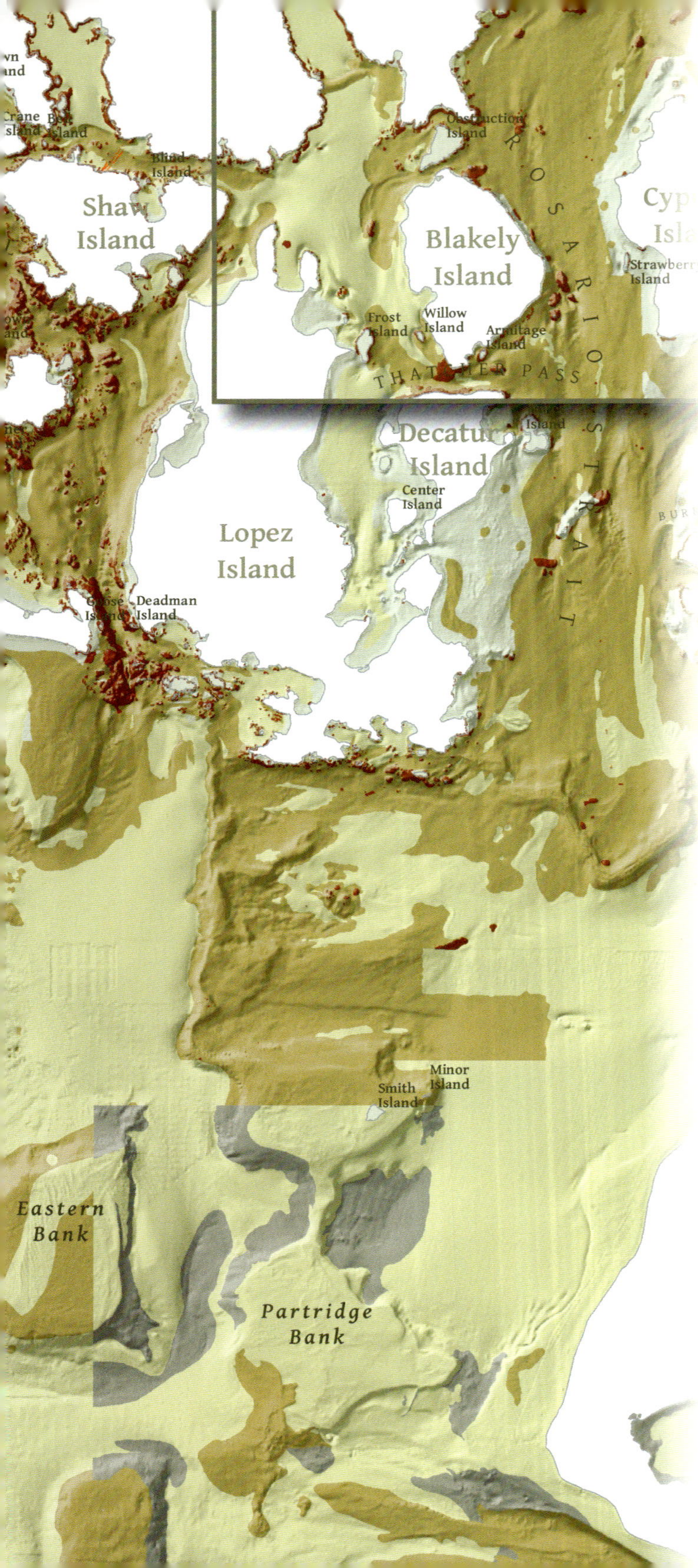

Substrate of Puget Sound »

CORE GIS

This map of the substrate of Puget Sound was created for The Nature Conservancy. TNC had just completed a two-year effort to compile data for use in marine spatial planning, and they wanted to produce a series of maps to showcase and promote the data. They asked me to design four maps showing bathymetry, substrate, geomorphology, and habitat types. The intended audience was primarily the marine research community, but they also wanted to be able to convey this information to the broader public, so it was important for the maps to be both technically engaging and aesthetically pleasing.

Mapping Puget Sound can be challenging—it has an awkward shape that can result in a lot of unnecessary white space. Fortunately, TNC was interested in highlighting several areas throughout Puget Sound that best convey a sense of the rich detail in the data. We identified these areas collaboratively and settled on nine inset maps distributed throughout the Sound. I arranged the inset maps in a rigid geometrical pattern, forming a visual counterpoint to the seemingly randomly distributed inset locations and the natural contours of the shorelines and islands. I used the remaining white space to place the extensive text that TNC provided.

The color palette is desaturated to make the annotation and inset locators stand out. There is little information in the terrestrial portion of the region, in order to focus the viewer on the marine features. There are seven categories of substrate shown on the map, which are depicted with muted colors. The exceptions are the three classes with the smallest spatial extent: rock, anthropogenic, and shell substrates. The colors for these features are more saturated, and, in the case of the first two, use warmer hues. My objective in making these color choices was to introduce visual interest and lead the map reader through an exploration of the surprising natural variety to be found beneath the surface of Puget Sound.

Substrate *of* Puget Sound

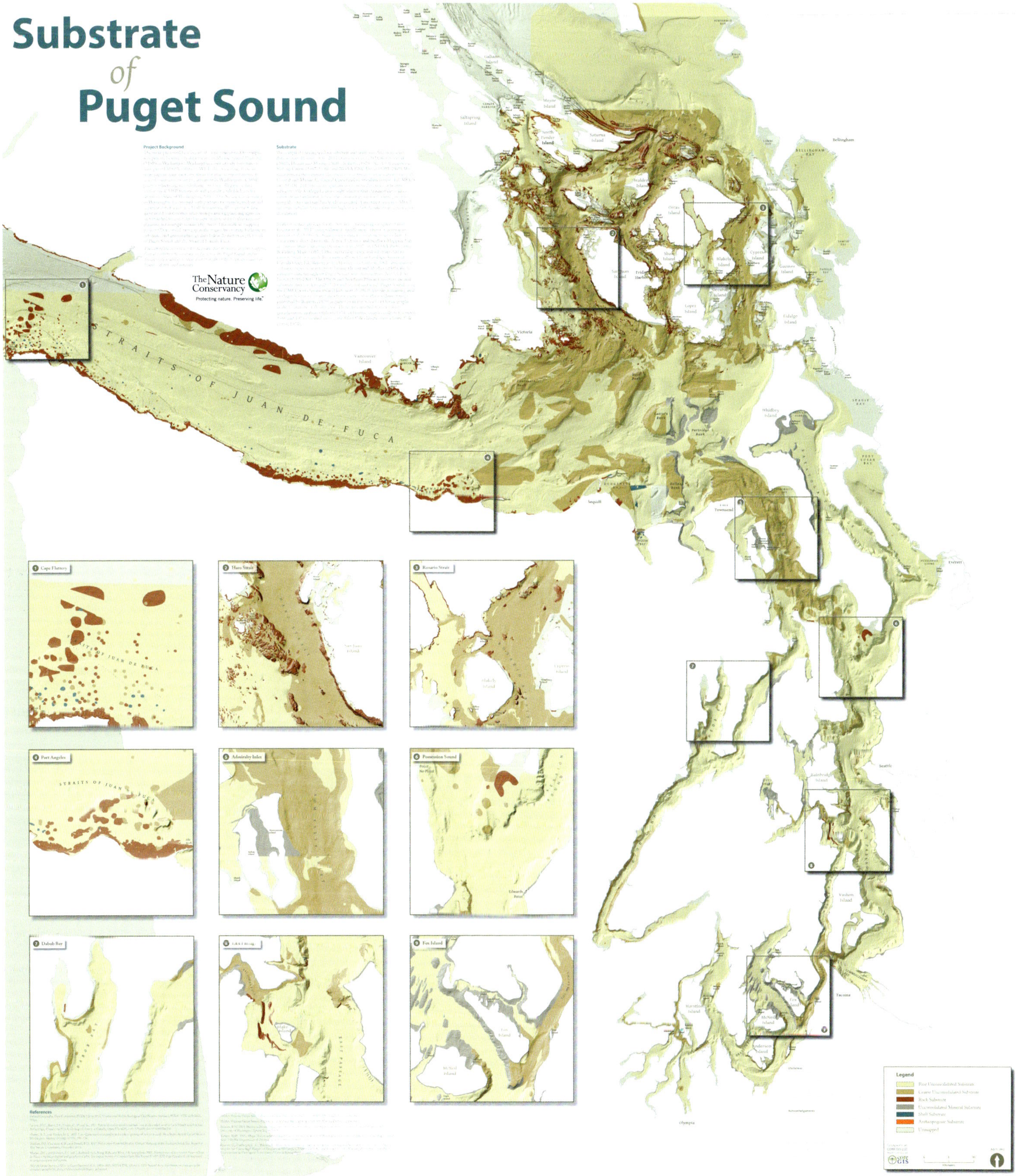

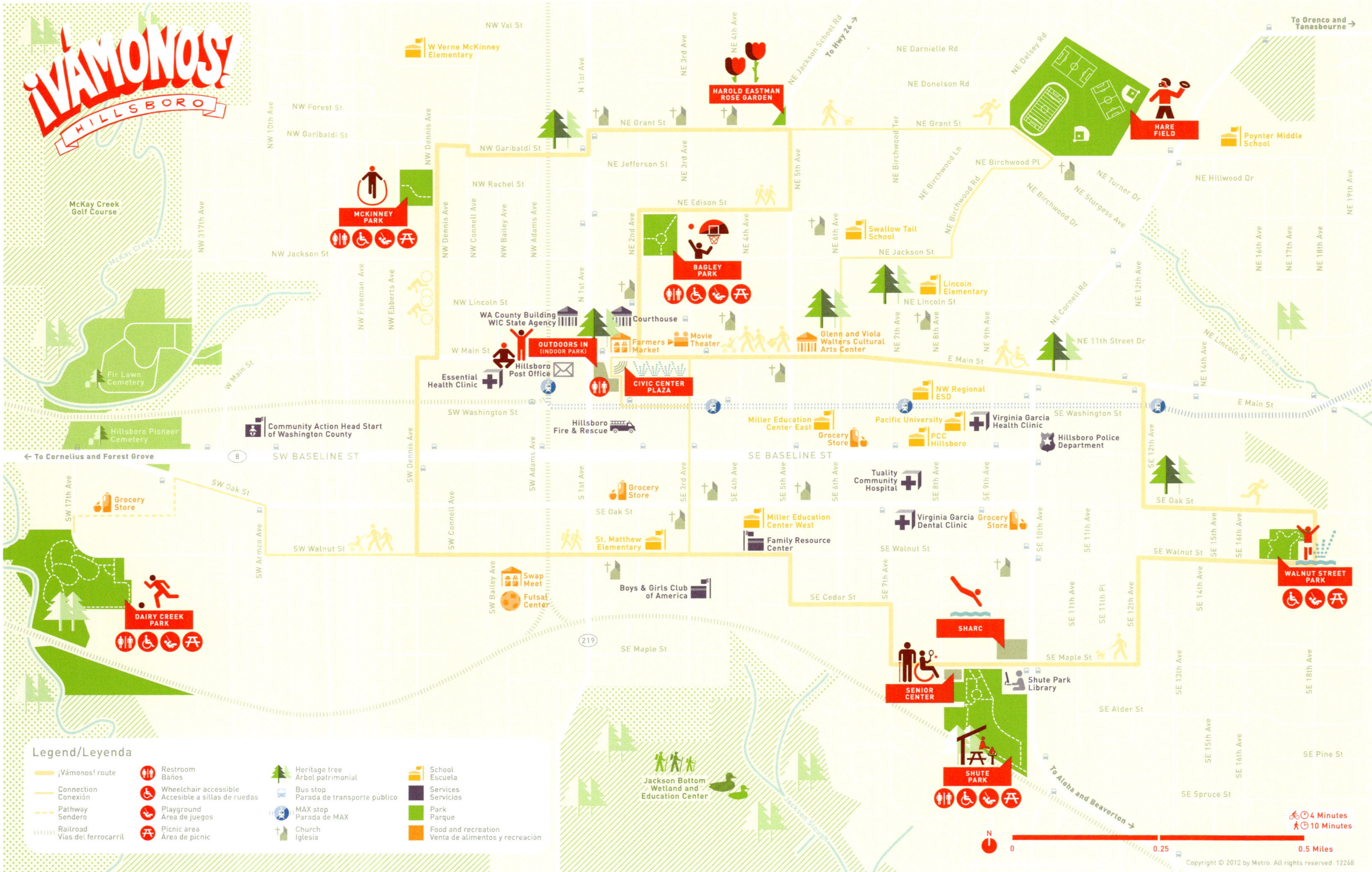

¡VÁMONOS!
HILLSBORO
W Verne McKinney Elementary
Harold Eastman Rose Garden
Hare Field
Poynter Middle School
To Orenco and Tanasbourne
To Hwy 26
McKinney Park
McKay Creek Golf Course
Bagley Park
Swallow Tail School
Lincoln Elementary
WA County Building
WIC State Agency
Courthouse
Outdoors in (indoor park)
Farmers Market
Movie Theater
Glenn and Viola Walters Cultural Arts Center
Hillsboro Post Office
Essential Health Clinic
Civic Center Plaza
NW Regional ESD
Fir Lawn Cemetery
Hillsboro Pioneer Cemetery
Community Action Head Start of Washington County
Hillsboro Fire & Rescue
Miller Education Center East
Pacific University
Virginia Garcia Health Clinic
Grocery Store
PCC Hillsboro
Hillsboro Police Department
To Cornelius and Forest Grove
SW BASELINE ST
SE BASELINE ST
Tuality Community Hospital
Grocery Store
Miller Education Center West
Virginia Garcia Dental Clinic
Grocery Store
St. Matthew Elementary
Family Resource Center
Walnut Street Park
Dairy Creek Park
Swap Meet
Futsal Center
Boys & Girls Club of America
SHARC
Senior Center
Shute Park Library
Shute Park
Jackson Bottom Wetland and Education Center
To Aloha and Beaverton
Legend/Leyenda
¡Vámonos! route
Connection / Conexión
Pathway / Sendero
Railroad / Vías del ferrocarril
Restroom / Baños
Wheelchair accessible / Accesible a sillas de ruedas
Playground / Área de juegos
Picnic area / Área de picnic
Heritage tree / Árbol patrimonial
Bus stop / Parada de transporte público
MAX stop / Parada de MAX
Church / Iglesia
School / Escuela
Services / Servicios
Park / Parque
Food and recreation / Venta de alimentos y recreación
4 Minutes
10 Minutes
0
0.25
0.5 Miles

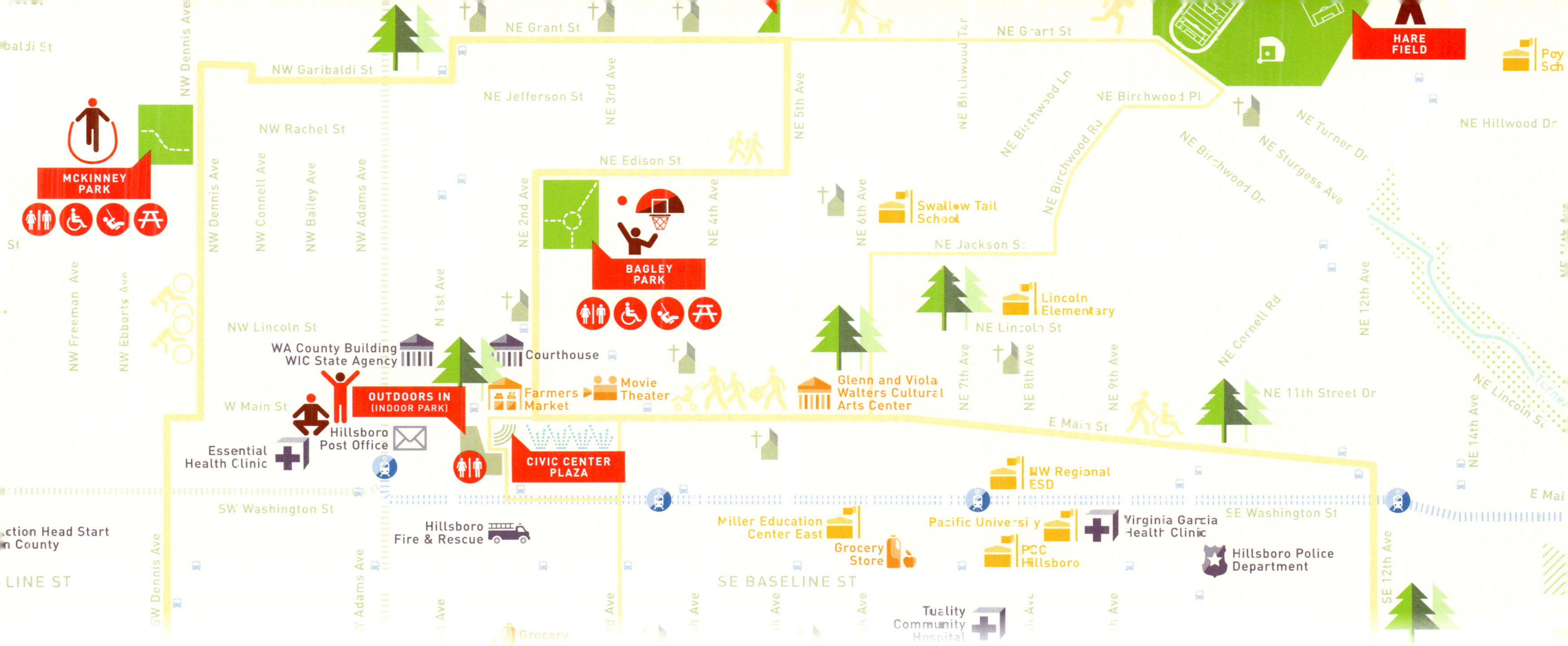

« ¡Vámonos!

Ryan **Sullivan** and Matthew **Hampton**

This set of bilingual walking maps promotes health and cultural awareness in Hillsboro, Cornelius, and Forest Grove, Oregon—three auto-oriented suburban towns with a large Hispanic population, located along the western edge of the Portland region's urban growth boundary. The maps illustrate great places to walk and bike, including cultural points of interest and recreational amenities, and are geared towards Latino families—the largest and fastest-growing minority in the Portland metropolitan area. Funded through a Kaiser grant, one objective of the project is to encourage healthy and sustainable modes of transportation by encouraging walking and physical activity. Over 60,000 maps were distributed via a summer-long program of associated outreach events.

The design strikes a careful balance in order to be colorful and playful without compromising legibility. In order to achieve this balance, legibility is addressed by organizing the content on a rigid grid and using a clear visual hierarchy, while the playfulness of the maps is expressed in their colors and the illustrative pictograms. We began the project by exploring a variety of case studies for inspiration—including Lance Wyman's graphic identity system for the 1968 Olympic Games in Mexico City and the work of Alexander Girard. Exported files from GIS software were redrawn in order to conform to a grid so that content could be organized consistently and simplified to a level of detail that was appropriate for the scale of the map. The design process continued with an exploration of many different color options and alternative concepts for the pictograms. Decisions about color and the effectiveness of the pictograms were guided by a focus-group process with the community.

» *We aimed to give the map a historical feel in keeping with the subject matter, while making sure that all historical embellishments were relevant and presented as "data."*

Defining an Empire »

Virginia W. ***Mason****, Matthew* ***Twombly****, and Amanda* ***Hobbs*** *— National Geographic Magazine*

Unlike many maps of the Roman Empire, this one illustrates both its physical and natural barriers. The Roman Empire wasn't delineated by a hard-defined border in many places. Rather, the map represents the extent of a frontier that depended on both the political and religious influences of the time and the constraints that natural barriers placed on travel. Because of differing interpretations of the historical record, we chose the input of one scholarly expert to define this frontier. His work focuses on the mid-second century, a time in history when Roman expansion was at its greatest.

We used hand-drawn cartographic symbols to highlight the main natural and constructed barriers that defined the Empire in the middle of the second century. Hill and mountain symbols were drawn using a digital elevation model as a reference. The mountain ranges and deserts that define the boundaries are emphasized. Careful thought was put into the presentation of these natural features. We aimed to give the map a historical feel in keeping with the subject matter, while making sure that all historical embellishments were relevant and presented as "data." For example, the subtle water texture indicates a water barrier; the destructed earth tones reveal desert barriers.

This map accompanied a *National Geographic* magazine story that features landscape photography of the world's best-preserved Roman forts. Romans built several types of fortifications, which are illustrated on the map. These illustrations were extensively researched, based on archeological and sociological information.

From a visual perspective, we didn't want the map and illustrations to compete with the photographs; we didn't want to make them photo-realistic. Our aim was to have a spread map that gives the reader a nice "break" by coalescing the visuals in the photography and the text in the manuscript. As a result, the spread tells its own story, indicating the Roman presence on a vast landscape.

DEFINING AN EMPIRE

By the middle of the second century, as depicted on this map, the Roman Empire's centuries of relentless expansion had slowed and then stopped. Rome relied on its network of forts, walls, and natural barriers to separate the empire from barbarians, the term for everyone outside its borders. A mixture of diplomacy, trade, and violence was used to keep the frontiers intact. Eventually, barbarian incursions led to the fall of the western part of the empire in the fifth century.

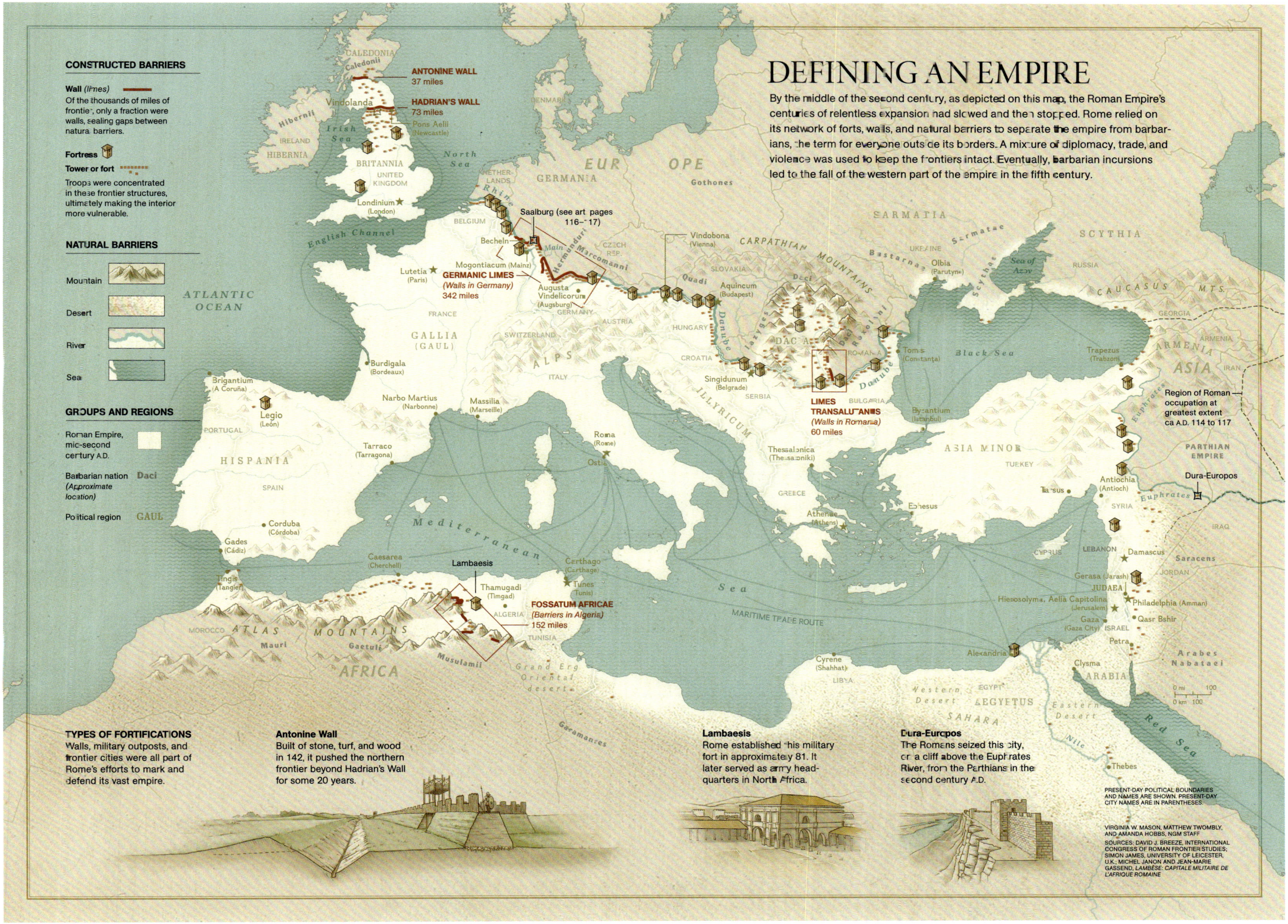

TYPES OF FORTIFICATIONS
Walls, military outposts, and frontier cities were all part of Rome's efforts to mark and defend its vast empire.

Antonine Wall
Built of stone, turf, and wood in 142, it pushed the northern frontier beyond Hadrian's Wall for some 20 years.

Lambaesis
Rome established this military fort in approximately 81. It later served as army headquarters in North Africa.

Dura-Europos
The Romans seized this city, on a cliff above the Euphrates River, from the Parthians in the second century A.D.

PRESENT-DAY POLITICAL BOUNDARIES AND NAMES ARE SHOWN. PRESENT-DAY CITY NAMES ARE IN PARENTHESES.

VIRGINIA W. MASON, MATTHEW TWOMBLY, AND AMANDA HOBBS, NGM STAFF

SOURCES: DAVID J. BREEZE, INTERNATIONAL CONGRESS OF ROMAN FRONTIER STUDIES; SIMON JAMES, UNIVERSITY OF LEICESTER, U.K.; MICHEL JANON AND JEAN-MARIE GASSEND, *LAMBÈSE: CAPITALE MILITAIRE DE L'AFRIQUE ROMAINE*

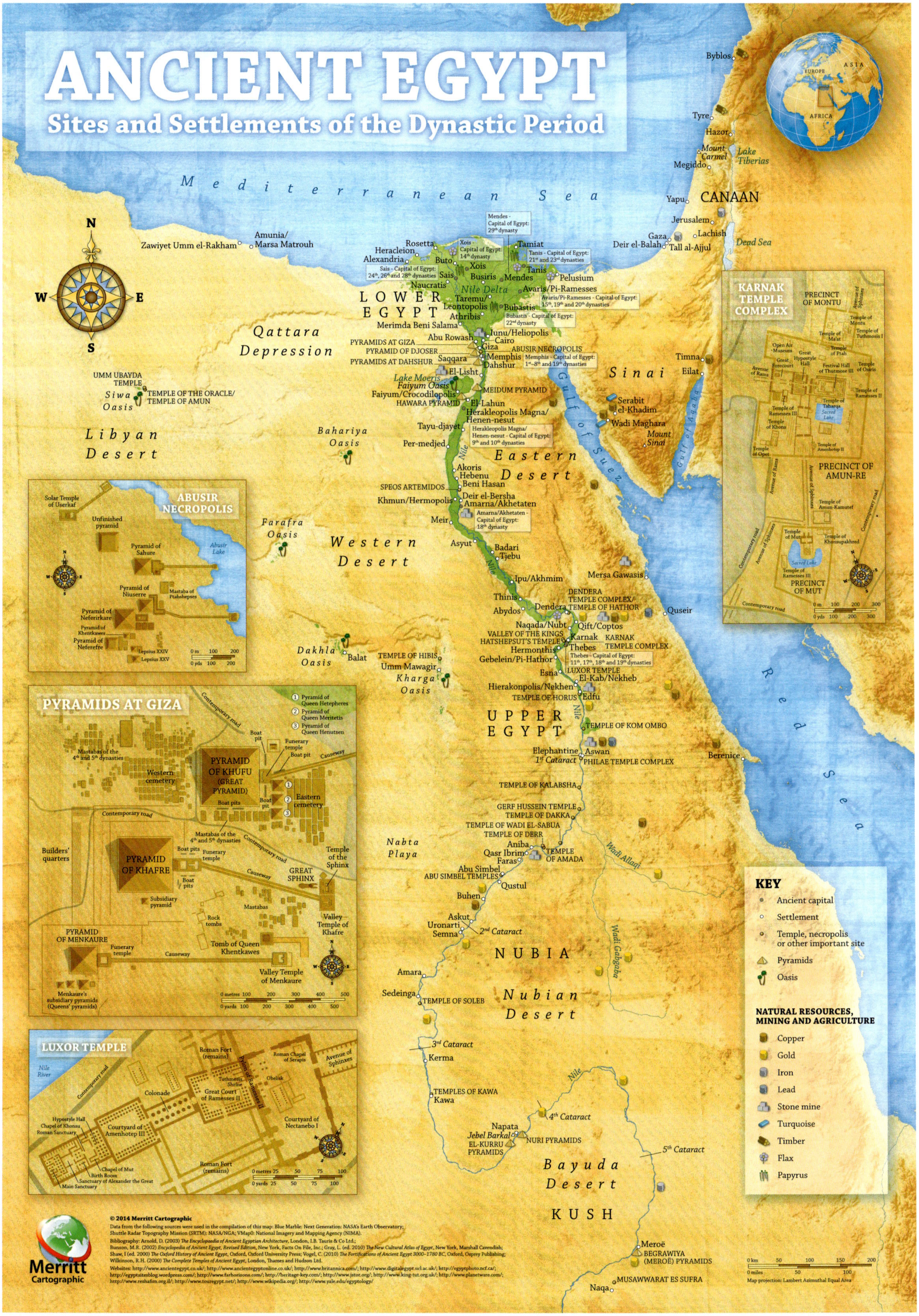

ANCIENT EGYPT
Sites and Settlements of the Dynastic Period
Mediterranean Sea
Red Sea
Gulf of Suez
Gulf of Aqaba
Dead Sea
Lake Tiberias
CANAAN
Byblos
Tyre
Hazor
Mount Carmel
Megiddo
Yapu
Jerusalem
Lachish
Gaza
Tall al-Ajjul
Deir el-Balah
Zawiyet Umm el-Rakham
Amunia/ Marsa Matrouh
Rosetta
Heracleion
Alexandria
Buto
Xois
Sais
Naucratis
Busiris
Mendes
Tamiat
Tanis
Pelusium
Avaris/Pi-Ramesses
Taremu/ Leontopolis
Bubastis
Athribis
Nile Delta
Mendes - Capital of Egypt: 29th dynasty
Xois - Capital of Egypt: 14th dynasty
Sais - Capital of Egypt: 24th, 26th and 28th dynasties
Tanis - Capital of Egypt: 21st and 23rd dynasties
Avaris/Pi-Ramesses - Capital of Egypt: 15th, 19th and 20th dynasties
Bubastis - Capital of Egypt: 22nd dynasty
LOWER EGYPT
Merimda Beni Salama
Abu Rowash
Junu/Heliopolis
Cairo
Giza
PYRAMIDS AT GIZA
PYRAMID OF DJOSER
PYRAMIDS AT DAHSHUR
ABUSIR NECROPOLIS
Saqqara
Memphis
Dahshur
Memphis - Capital of Egypt: 1st–8th and 13th dynasties
El-Lisht
Lake Moeris
Faiyum Oasis
Faiyum/Crocodilopolis
HAWARA PYRAMID
MEIDUM PYRAMID
El-Lahun
Herakleopolis Magna/ Henen-nesut
Herakleopolis Magna/ Henen-nesut - Capital of Egypt: 9th and 10th dynasties
Tayu-djayet
Per-medjed
Qattara Depression
UMM UBAYDA TEMPLE
TEMPLE OF THE ORACLE/ TEMPLE OF AMUN
Siwa Oasis
Libyan Desert
Bahariya Oasis
Farafra Oasis
Western Desert
Eastern Desert
Sinai
Timna
Eilat
Serabit el-Khadim
Wadi Maghara
Mount Sinai
Akoris
Hebenu
Beni Hasan
SPEOS ARTEMIDOS
Khmun/Hermopolis
Deir el-Bersha
Amarna/Akhetaten
Amarna/Akhetaten - Capital of Egypt: 18th dynasty
Meir
Asyut
Badari
Tjebu
Ipu/Akhmim
Mersa Gawasis
Thinis
Abydos
Dendera
DENDERA TEMPLE COMPLEX/ TEMPLE OF HATHOR
Quseir
Naqada/Nubt
Qift/Coptos
VALLEY OF THE KINGS
HATSHEPSUT'S TEMPLE
Karnak
KARNAK TEMPLE COMPLEX
Hermonthis
Thebes
Thebes - Capital of Egypt: 11th, 17th, 18th and 19th dynasties
Gebelein/Pi-Hathor
LUXOR TEMPLE
Esna
El-Kab/Nekheb
Hierakonpolis/Nekhen
TEMPLE OF HORUS
Edfu
Dakhla Oasis
Balat
TEMPLE OF HIBIS
Umm Mawagir
Kharga Oasis
UPPER EGYPT
Nile
TEMPLE OF KOM OMBO
Elephantine
Aswan
1st Cataract
PHILAE TEMPLE COMPLEX
Berenice
TEMPLE OF KALABSHA
GERF HUSSEIN TEMPLE
TEMPLE OF DAKKA
TEMPLE OF WADI EL-SABUA
TEMPLE OF DERR
Aniba
Qasr Ibrim
TEMPLE OF AMADA
Faras
Abu Simbel
ABU SIMBEL TEMPLES
Qustul
Nabta Playa
Wadi Allaqi
Wadi Gabgaba
Buhen
Askut
Uronarti
Semna
2nd Cataract
NUBIA
Nubian Desert
Amara
Sedeinga
TEMPLE OF SOLEB
3rd Cataract
Kerma
TEMPLES OF KAWA
Kawa
4th Cataract
Napata
Jebel Barkal
EL-KURRU PYRAMIDS
NURI PYRAMIDS
Bayuda Desert
5th Cataract
KUSH
Meroë
BEGRAWIYA (MEROË) PYRAMIDS
MUSAWWARAT ES SUFRA
Naqa
N
S
E
W
EUROPE
ASIA
AFRICA
KARNAK TEMPLE COMPLEX
PRECINCT OF MONTU
Avenue of Sphinxes
Temple of Montu
Temple of Ma'at
Temple of Tuthmosis I
Open Air Museum
Great Forecourt
Great Hypostyle Hall
Temple of Ptah
Festival Hall of Thutmose III
Temple of Osiris
Avenue of Rams
Temple of Ramesses III
Temple of Taharqa
Sacred Lake
Temple of Ramesses II
Temple of Khonsu
Temple of Opet
Temple of Amenhotep II
PRECINCT OF AMUN-RE
Temple of Amun-Kamutef
Temple of Mut
Temple of Khonspakhred
Temple of Ramesses III
PRECINCT OF MUT
Contemporary road
0 m 100 200 300
0 yds 100 200 300
ABUSIR NECROPOLIS
Solar Temple of Userkaf
Unfinished pyramid
Pyramid of Sahure
Abusir Lake
Pyramid of Niuserre
Mastaba of Ptahshepses
Pyramid of Neferirkare
Pyramid of Khentkawes
Pyramid of Neferefre
Lepsius XXIV
Lepsius XXV
0 m 100 200
0 yds 100 200
PYRAMIDS AT GIZA
1 Pyramid of Queen Hetepheres
2 Pyramid of Queen Meritetis
3 Pyramid of Queen Henutsen
Contemporary road
Boat pit
Funerary temple
Causeway
Mastabas of the 4th and 5th dynasties
Western cemetery
PYRAMID OF KHUFU (GREAT PYRAMID)
Boat pits
Eastern cemetery
Builders' quarters
PYRAMID OF KHAFRE
Subsidiary pyramid
Rock tombs
Mastabas
Temple of the Sphinx
GREAT SPHINX
Valley Temple of Khafre
Tomb of Queen Khentkawes
PYRAMID OF MENKAURE
Menkaure's subsidiary pyramids (Queens' pyramids)
Valley Temple of Menkaure
0 metres 100 200 300 400 500
0 yards 100 200 300 400 500
LUXOR TEMPLE
Nile River
Contemporary road
Roman Fort (remains)
Roman Chapel of Serapis
Avenue of Sphinxes
Pylon of Ramesses II
Obelisk
Colonade
Great Court of Ramesses II
Courtyard of Nectanebo I
Hypostyle Hall
Chapel of Khonsu
Roman Sanctuary
Courtyard of Amenhotep III
Chapel of Mut
Birth Room
Sanctuary of Alexander the Great
Main Sanctuary
0 metres 25 50 75 100
0 yards 25 50 75 100
KEY
Ancient capital
Settlement
Temple, necropolis or other important site
Pyramids
Oasis
NATURAL RESOURCES, MINING AND AGRICULTURE
Copper
Gold
Iron
Lead
Stone mine
Turquoise
Timber
Flax
Papyrus
0 km 50 100 150 200
0 miles 50 100
Map projection: Lambert Azimuthal Equal Area
Merritt Cartographic
© 2014 Merritt Cartographic
Data from the following sources were used in the compilation of this map: Blue Marble: Next Generation: NASA's Earth Observatory; Shuttle Radar Topography Mission (SRTM): NASA/NGA; VMap0: National Imagery and Mapping Agency (NIMA).
Bibliography: Arnold, D. (2003) The Encyclopaedia of Ancient Egyptian Architecture, London, I.B. Tauris & Co Ltd.; Bunson, M.R. (2002) Encyclopedia of Ancient Egypt, Revised Edition, New York, Facts On File, Inc.; Gray, L. (ed. 2010) The New Cultural Atlas of Egypt, New York, Marshall Cavendish; Shaw, I (ed. 2000) The Oxford History of Ancient Egypt, Oxford, Oxford University Press; Vogel, C. (2010) The Fortifications of Ancient Egypt 3000–1780 BC, Oxford, Osprey Publishing; Wilkinson, R.H. (2000) The Complete Temples of Ancient Egypt, London, Thames and Hudson Ltd.
Websites: http://www.ancientegypt.co.uk/; http://www.ancientegyptonline.co.uk/; http://www.britannica.com/; http://www.digitalegypt.ucl.ac.uk/; http://egyptphoto.ncf.ca/; http://egyptsitesblog.wordpress.com/; http://www.farhorizons.com/; http://heritage-key.com/; http://www.jstor.org/; http://www.king-tut.org.uk/; http://www.planetware.com/; http://www.reshafim.org.il/; http://www.touregypt.net/; http://www.wikipedia.org/; http://www.yale.edu/egyptology/

« Ancient Egypt: Sites and Settlements of the Dynastic Period

Ed **Merritt**

This is not a commissioned piece. I wanted to learn more about ancient Egypt, and I have rarely, if ever, found a better way to study a subject than to create a map about it. Despite the existence of numerous maps which already cover this topic, I knew that, given the timeless popularity and intrigue surrounding the ancient Egyptian civilisation, I would not be lacking an audience.

I have attempted to incorporate several characteristics of ancient Egyptian art into my design. The use of strong and sometimes vibrant colours, derived primarily from mineral compounds and able to resist the effects of powerful sunlight, is a trait of ancient Egyptian paintings. It seemed appropriate, therefore, to use reasonably saturated colours in my design. Slab-serif fonts have been used throughout. Although these do not have any direct connections with Egyptian writing systems, fonts in this form are typically employed on contemporary products relating to ancient Egypt, and I was happy to follow that trend here. I used a relatively heavy texture in the background to give the map a slightly weathered appearance, emphasising the amount of time that has passed between now and the subject being mapped. This texture is derived from a photograph of a papyrus sheet, a product used by ancient Egyptians as a writing material. Children frequently encounter ancient Egypt in their studies, and I wanted to recognise this in my design by including a range of illustrative symbols noting important mining and agricultural locations, which may appeal to the younger reader.

The process of creating this map involved seeking out as many resources as I could find, including books, old maps, and information on websites, before I began collating and drawing together as much relevant detail as I could obtain from these resources. It would not have been possible to produce this piece without the efforts of all those authors and researchers whose materials I referenced, and so I am very grateful for their help.

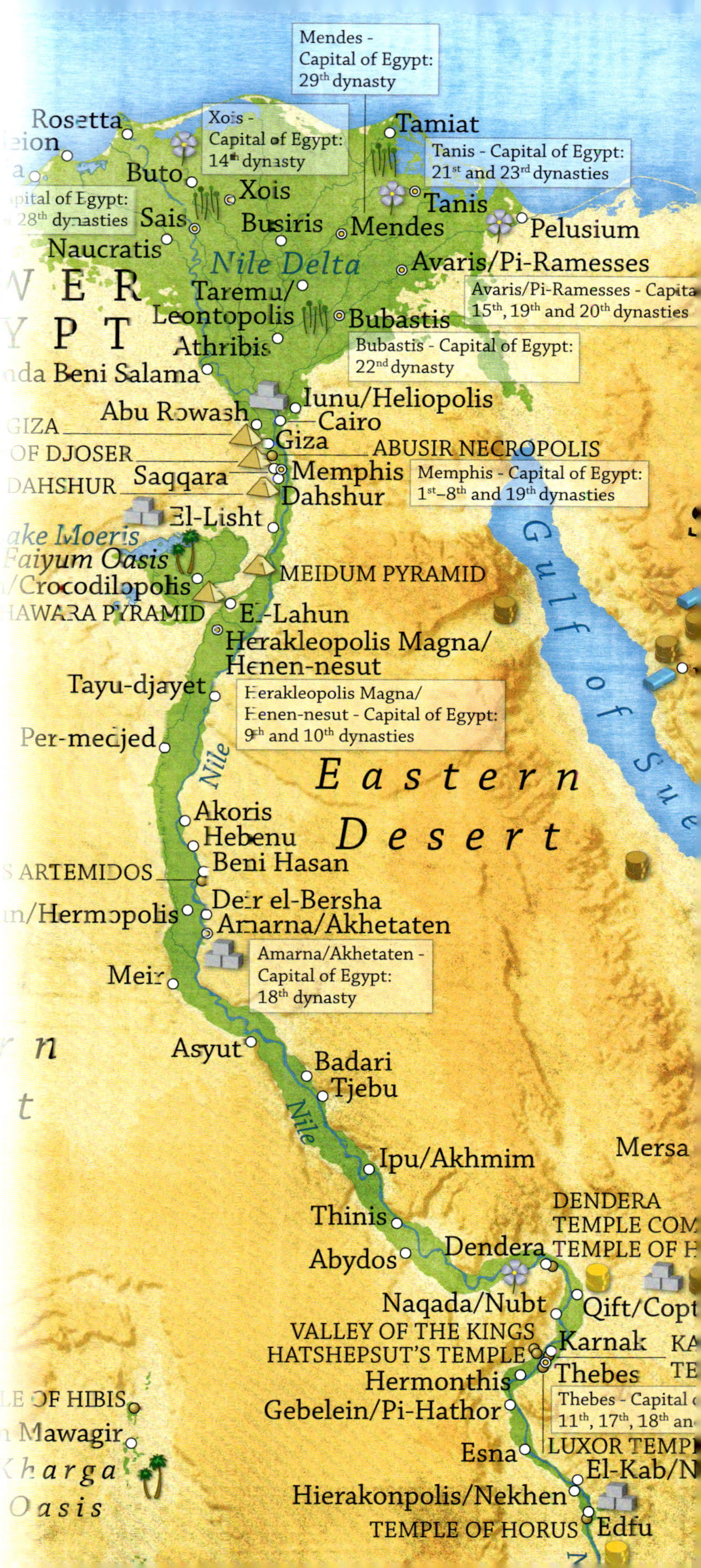

Australia's Deserts: a Köppen Climate Map of the Australian Mainland »

Dan **Bowles**

» *"What is a desert?" The answer: it depends.*

This map of the Australian mainland was created for a special deserts issue of *Australian Geographic*, to depict that nation's deserts and their human exploration. Key to this was asking the question, "What is a desert?" The answer: it depends. Deserts can be named places which have defined boundaries, shown on this map outlined in grey. These are human constructs and somewhat arbitrary.

Scientifically speaking, however, deserts are defined by various classification systems, the most prominent being that of Russian-German climatologist Wladimir Köppen dating from 1884. It designates a range of climate zones based primarily on temperatures and precipitation as they relate to native vegetation.

I wanted to depict the Köppen system through colour, and explicitly show the reader the regions of the Australian continent that it designated as desert, which I outlined in a reddish-brown. To do this, I created a bivariate map depicting both annual mean temperature and precipitation. Precipitation is depicted across a brown to green colour ramp and temperature across a blue-white-red colour ramp. Desert and arid regions show a rusty brown, the forests of the southeast a cool, lush green, and the tropical savannahs of the north a warmer green-brown. Colours on the map combine to form an intuitive portrayal of Australia's climate.

The map displays bivariate data using several innovative methods. The majority of bivariate maps are choropleth maps, meaning they display data as discrete blocks of colour. Temperature and precipitation averages, however, are continuous surfaces and therefore climate zones meld into one another; they do not generally have distinct boundaries. For this reason, I decided to use a continuous colour surface to represent climate. The second innovative feature of the map is in the colour ramps themselves. Most bivariate maps use colour saturation of two individual hues to create the colour space. This map uses four hues, two for each variant, to create a much more complex and nuanced colour space.

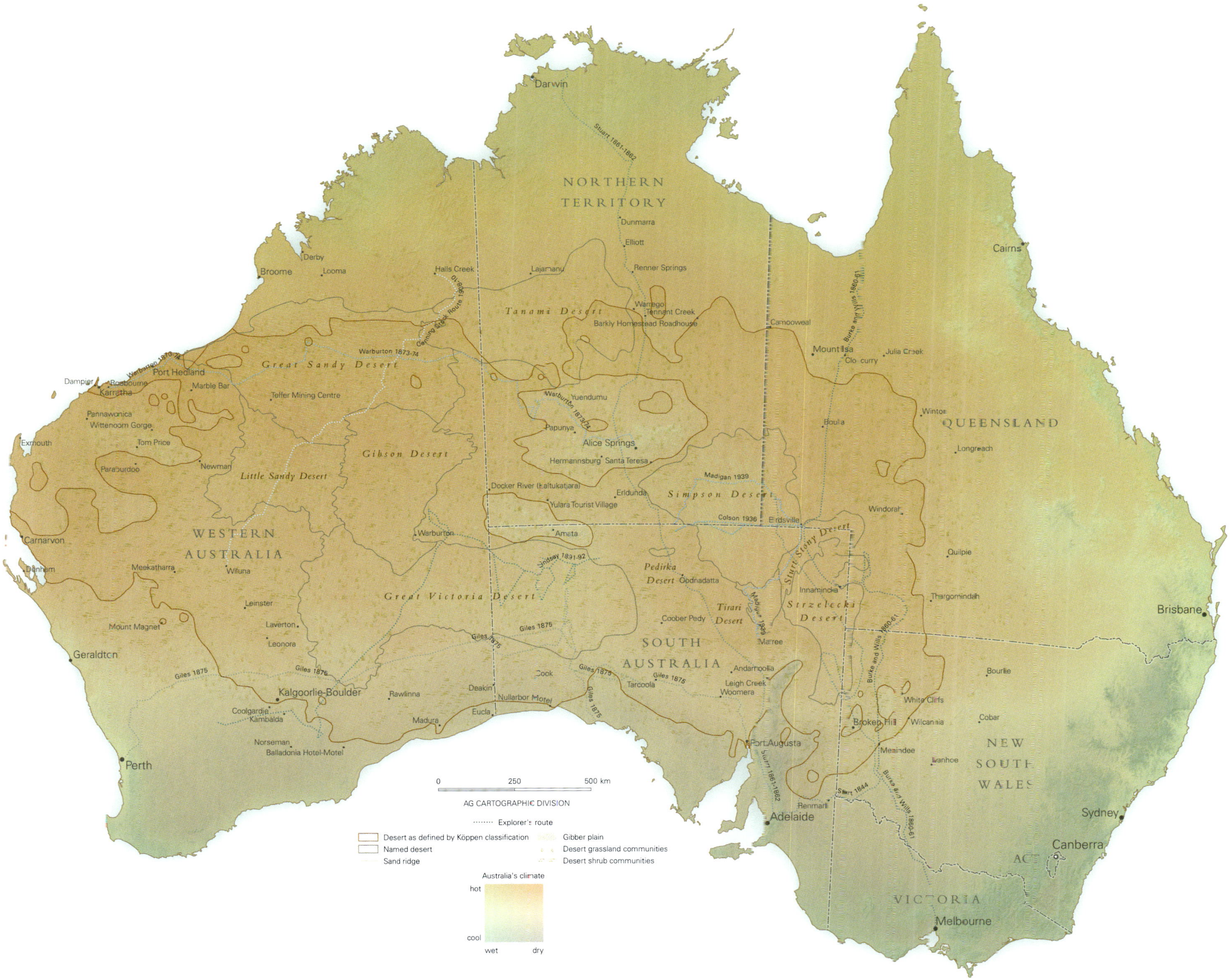

Darwin
Stuart 1861-1862
NORTHERN
TERRITORY
Dunmarra
Elliott
Renner Springs
Lajamanu
Halls Creek
Derby
Broome
Looma
Canning Stock Route 1906-10
Tanami Desert
Warrego
Tennant Creek
Barkly Homestead Roadhouse
Camooweal
Burke and Wills 1860-61
Mount Isa
Julia Creek
Cloncurry
Cairns
Warburton 1873-74
Great Sandy Desert
Port Hedland
Dampier
Roebourne
Karratha
Marble Bar
Telfer Mining Centre
Yuendumu
Papunya
Alice Springs
Hermannsburg
Santa Teresa
Boulia
Winton
QUEENSLAND
Longreach
Pannawonica
Wittenoom Gorge
Exmouth
Tom Price
Paraburdoo
Newman
Little Sandy Desert
Gibson Desert
Madigan 1939
Docker River (Kaltukatjara)
Erldunda
Simpson Desert
Yulara Tourist Village
Colson 1936
Birdsville
Windorah
Carnarvon
WESTERN
AUSTRALIA
Warburton
Amata
Sturt Stony Desert
Quilpie
Denham
Meekatharra
Wiluna
Lindsay 1891-92
Pedirka
Desert
Oodnadatta
Innamincka
Great Victoria Desert
Tirari
Desert
Madigan 1939
Strzelecki
Desert
Thargomindah
Leinster
Coober Pedy
Brisbane
Laverton
Mount Magnet
Leonora
Giles 1875
SOUTH
AUSTRALIA
Marree
Burke and Wills 1860-61
Geraldton
Giles 1875
Cook
Andamooka
Bourke
Tarcoola
Leigh Creek
Woomera
White Cliffs
Kalgoorlie-Boulder
Rawlinna
Deakin
Nullarbor Motel
Coolgardie
Kambalda
Eucla
Madura
Broken Hill
Wilcannia
Cobar
Norseman
Balladonia Hotel-Motel
Port Augusta
Menindee
Ivanhoe
NEW
SOUTH
WALES
Perth
Sturt 1861-1862
Sturt 1844
Renmark
0
250
500 km
AG CARTOGRAPHIC DIVISION
Sydney
Adelaide
Canberra
ACT
VICTORIA
Melbourne
Explorer's route
Desert as defined by Köppen classification
Named desert
Sand ridge
Gibber plain
Desert grassland communities
Desert shrub communities
Australia's climate
hot
cool
wet
dry

NATIONAL GEOGRAPHIC

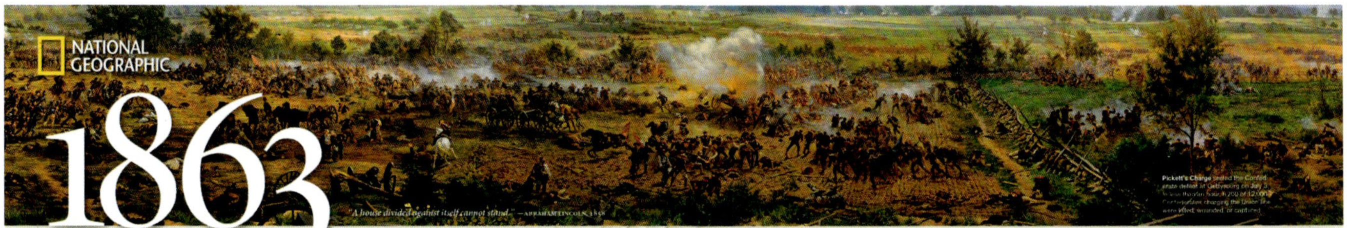

1863

TURNING POINT OF THE CIVIL WAR

From April 1861 to April 1865, the slaveholding states of the Confederacy fought for independence from the Union that President Abraham Lincoln was determined to preserve. Both sides had expected to win the war quickly. But the Union began to gain the upper hand only in July 1863 with its victory at Gettysburg, the largest battle ever fought in North America. The map at right tracks two opposing armies through that fateful year. Their march starts at the bottom of this poster.

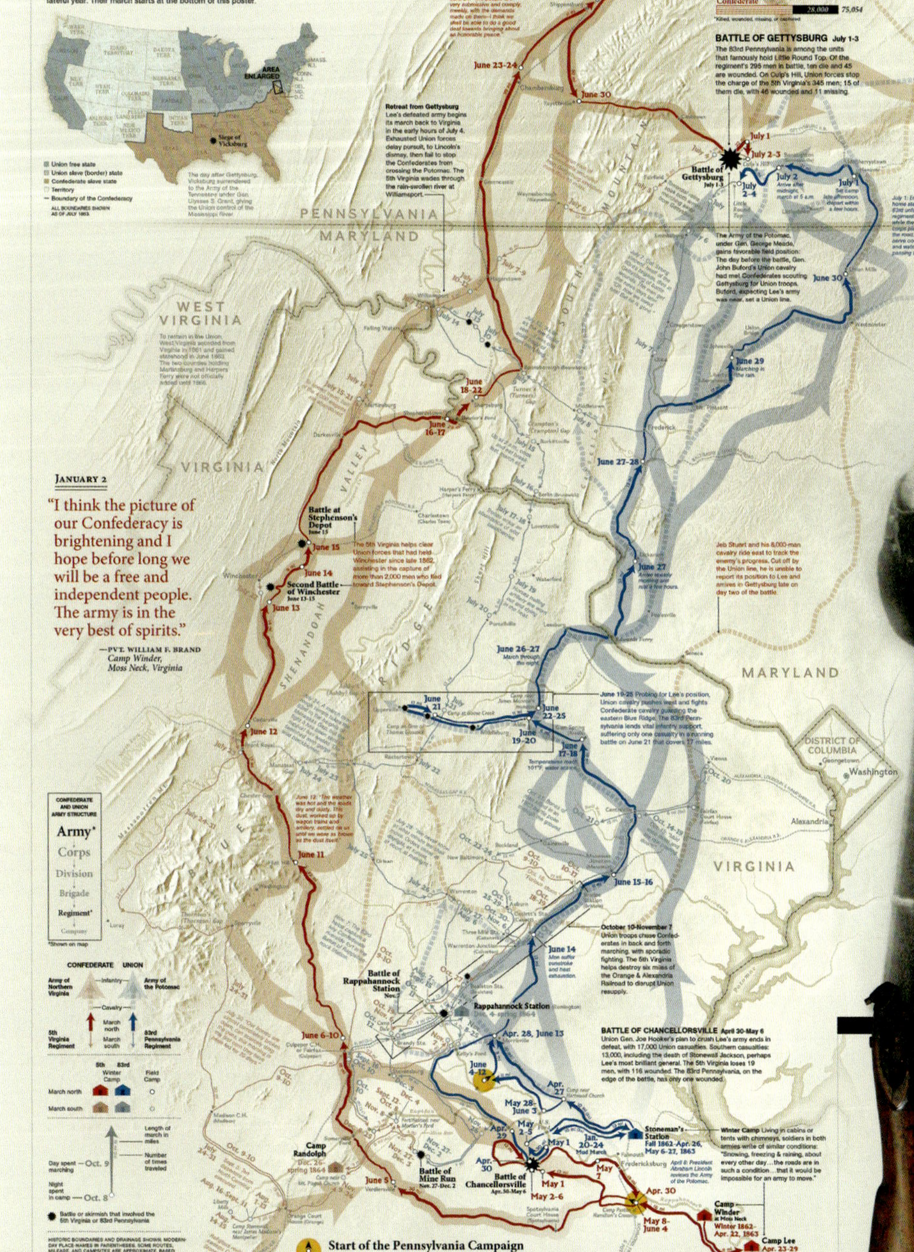

Start of the Pennsylvania Campaign

June 3, a month after the Battle of Chancellorsville, the 75,000-man Army of Northern Virginia heads north in stages, to draw less attention. The 5th Virginia marches on June 5. The Blue Ridge mountains shield the army's lightning sweep down the Shenandoah Valley. Union intelligence reports suggest Lee is moving, but the Army of the Potomac, 90,000 strong, does not move out until June 13.

JANUARY 2

"I think the picture of our Confederacy is brightening and I hope before long we will be a free and independent people. The army is in the very best of spirits."

—PVT. WILLIAM F. BRAND
Camp Winder, Moss Neck, Virginia

JULY 12

"Thousands of the men are barefoot… but they are bravely struggling on… so we may finish the war now… I am tired, almost worn out, haven't had my shoes off for a week, lying sometimes in the heaviest rain."

—PVT. OLIVER W. NORTON
Pursuing Confederates in retreat after Gettysburg, south of Hagerstown, Maryland

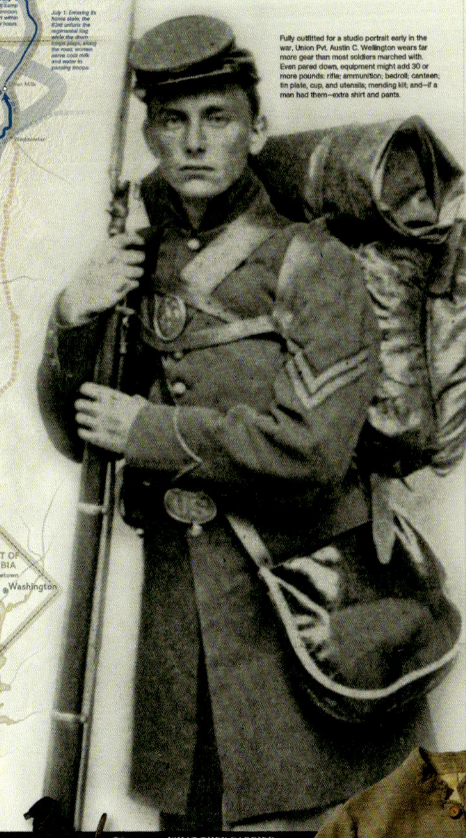

Fully outfitted for a studio portrait early in the war, Union Pvt. Austin C. Wellington wears far more gear than most soldiers marched with. Even pared down, equipment might add 30 or more pounds: rifle; ammunition; bedroll; canteen; tin plate, cup, and utensils; mending kit; and—if a man had them—extra shirt and pants.

WHAT THEY CARRIED

Battle was the least common event in a Civil War soldier's life. Mainly he faced the tedium of camp and marched, day or night, in mud, choking dust, heat, rain, or snow.

Wool uniforms were scratchy and often ill fitting. The North's advantage in manufacturing kept its men better outfitted. Of the South, Gen. Stonewall Jackson is said to have remarked: "My men have no uniform; they wear multiform."

Daily rations meant salt pork or beef, beans, and dried vegetables pressed into a cake. Northern troops relied on dense crackers called hardtack, often soaked in coffee for softening—and to drown the bugs. Corn bread was the Southern staple, judged most edible when fried in grease.

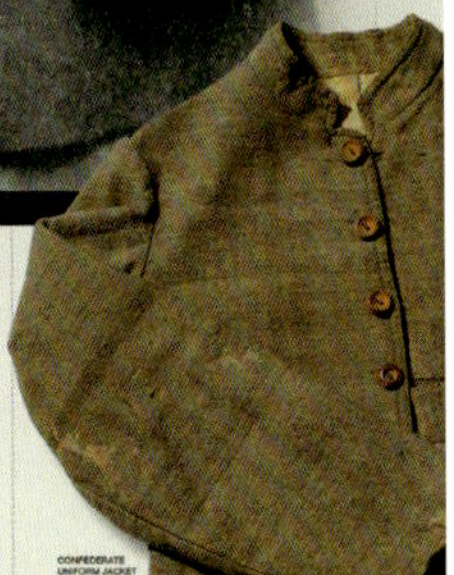

CONFEDERATE UNIFORM JACKET

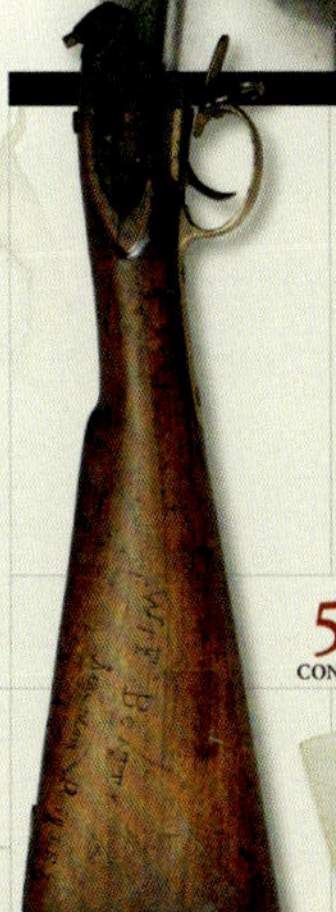

50% CONFEDERATE | 29% UNION ◀ A SOLDIER'S CHANCE OF BECOMING A CASUALTY*

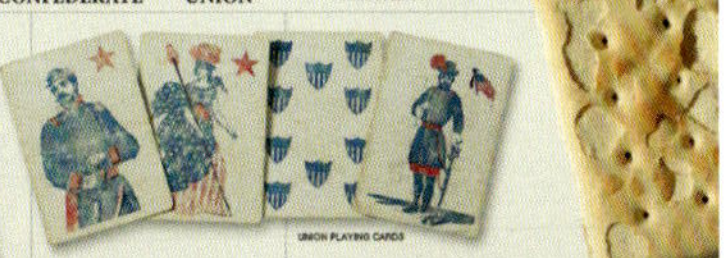

UNION PLAYING CARDS

Cards and other gambling games broke the boredom of camp.

Ankle-high boots called brogans usually fit poorly and wore out quickly. At many points in the war, thousands of Southern troops were barefoot.

Amputation saws were often the only option for bullet wounds: The .58-caliber lead minié ball lacerated tissue and shattered bone. Ether or chloroform, when available, sedated the patient. Germ-killing antiseptics were largely unknown, but nearly 75 percent of amputees survived.

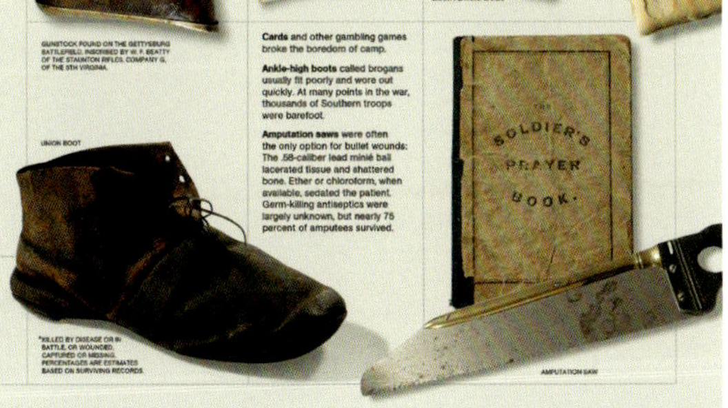

THE MARCH
TO GETTYSBURG AND BACK

5TH VIRGINIA INFANTRY
Army of Northern Virginia
Confederate States of America

732 MILES MARCHED IN 1863

This regiment of a thousand volunteers from the Shenandoah Valley was down to some 600 men by 1863. With four other regiments, it made up the famous Stonewall Brigade, honed by Gen. Thomas "Stonewall" Jackson. The previous year the regiment had marched 1,500 miles.

The Foot Soldier's War

General Robert E. Lee was hopeful in January 1863. His Confederate Army of Northern Virginia had taken unimaginable losses the previous year, but so had the Union's Army of the Potomac. The two forces were camped in Virginia on either side of the Rappahannock River to wait out the snowy, muddy winter. In May, Lee triumphed at the Battle of Chancellorsville. Now he aimed to invade Pennsylvania to get much needed food and to draw Union troops north into a battle he expected to win. It happened, by chance, at Gettysburg. Fought by more than 158,000 men, that pivotal battle went to the Union. This map focuses on two infantry regiments, the 5th Virginia and the 83rd Pennsylvania, tracing their march north and back through field reports, diaries, and letters home.

83RD PENNSYLVANIA INFANTRY
Army of the Potomac
United States of America

685 MILES MARCHED IN 1863

From the lakeside city of Erie and nearby farms, a thousand volunteers formed this regiment in 1861. Battle and disease had whittled it to some 400 men by 1863. Its bugler, Oliver Norton, was the first to play "Taps" as we know it, the tune arranged by his general, Daniel Butterfield.

1863: REGIMENT LIFE BY THE DAY

5TH Virginia: 124 days (Days spent in WINTER CAMPS) · 178 days FIELD CAMPS · 47 days MARCHES · 16 days BATTLES

83RD Pennsylvania: 13 days BATTLES · 62 days MARCHES · 130 days FIELD CAMPS · 160 days WINTER CAMPS

« The March to Gettysburg and Back: Turning Point of the Civil War

Maggie ***Smith***, *Bob* ***Pratt***, *Alexander* ***Stegmaier***, *Jane* ***Vessels***, *Amanda* ***Hobbs***, *Jeanne M.* ***Modderman***, *Bill* ***Marr***, *John* ***Tomanio***, *Mary Beth* ***Oelkers-Keegan***, *Theodore A.* ***Sickley***, *Rhett* ***Register***, *and David W.* ***Wooddell*** *— National Geographic Magazine*

A century and a half after the Battle of Gettysburg, it's hard to imagine thousands of soldiers trudging through the Blue Ridge Mountains toward an unknown destination. This map is intended to bring readers back to 1863, on an arduous and twisted year-long journey with Civil War troops as they marched to the Battle of Gettysburg and back. The team that designed this poster, a supplement to *National Geographic* magazine in May 2012, aimed for this map to be elegant, yet complex.

At first glance, this map contains a wealth of information, but an intentional hierarchy is clear. Thick arrows catch the eye, to give a quick sense of movement and context, showing readers that the Confederate army, in red, and the Union army, in blue, marched north toward Gettysburg on separate, winding paths. Thinner, yet more saturated and tapered arrows symbolize the two subject regiments, whose every mile is accounted for and meticulously labeled. The experiences of these soldiers are sprinkled throughout their journey, in the form of well-crafted map notes and larger pull quotes.

This map tells a history that had never before been fully pieced together. Its production thus required extensive research, working with expert historians and digging through archival maps, photographs, letters, and diaries containing hints to the whereabouts of these soldiers for every day of 1863.

Overall, we wanted the poster to express a historical aesthetic similar to its source materials. Water vignettes were carefully polished, and earthy tones chosen for base map lines and labels. Saturation and value were employed to draw attention to important features, such as winter camps, battle symbols, regiment arrows, and journey starting points.

The physical landscape played a big role in the lives of these soldiers, as they faced steep mountain passes and forded flooded rivers. The shaded relief is thus sharp and realistic, and drainage was edited for historical accuracy. We hope this map appeals to a broader audience, while satisfying the desires of Civil War aficionados who crave the minute details we painstakingly included.

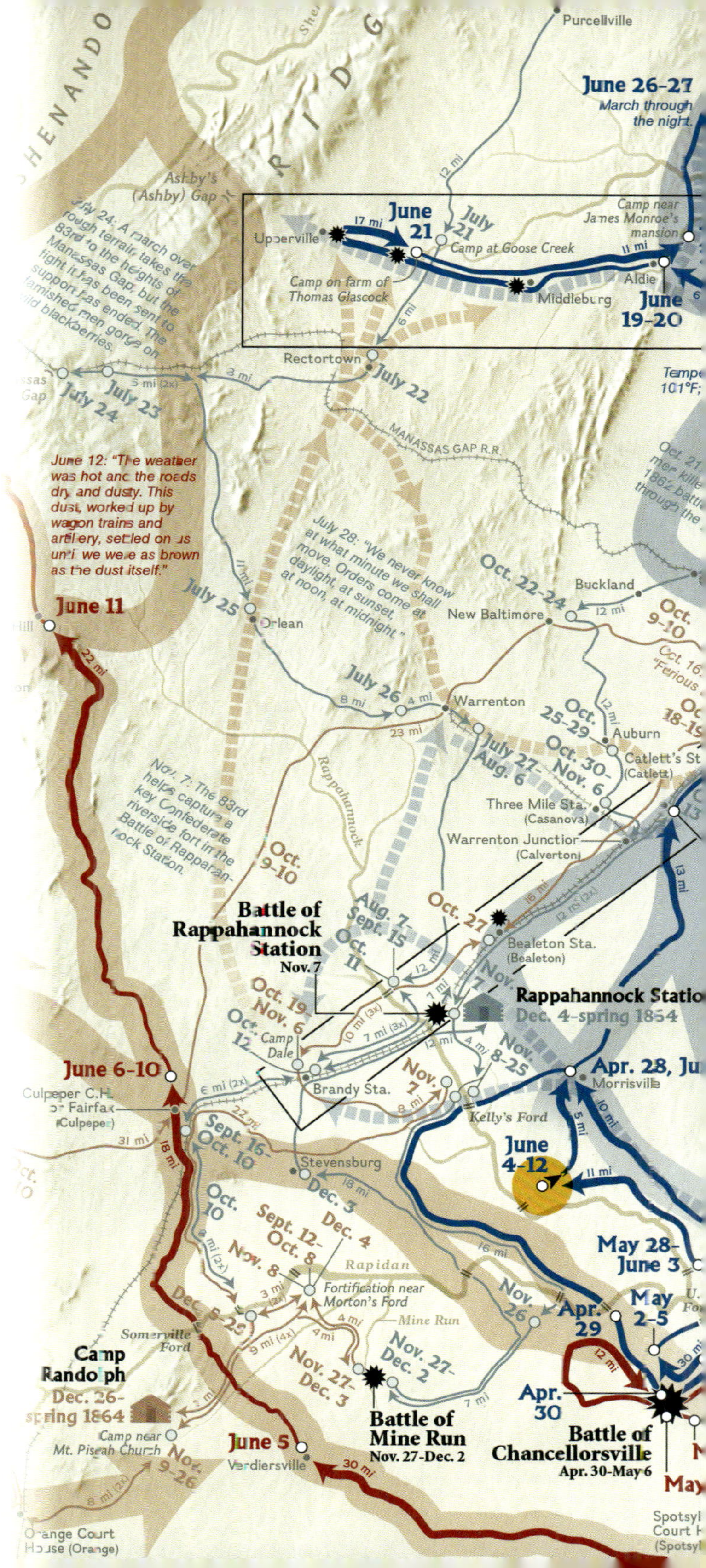

Collision Course »

Martin **Gamache** — *National Geographic Magazine*

» *Rarely as a* National Geographic *cartographer does a story affect me the way this one continues to do.*

On May 31, 2013, a powerful, erratic tornado killed a *National Geographic* storm chasing team in El Reno, Oklahoma. The victims included Tim Samaras, one of the most experienced and cautious storm chasers. This map illustrates the last 20 minutes of the team's chase and consequently Tim's and his chase partners' (Tim's son Paul and colleague Carl Young) lives. Rarely as a *National Geographic* cartographer does a story affect me the way this one continues to do.

The map you see here is the distilled analysis of the storm's progress as it moved through the Oklahoma farmland, taking an unexpected turn and jumping in size as it increased its ground speed minutes before tearing into three of our own. The design is understated, the palette subdued. I use the natural grid of the rural road network and the tornado sizes to communicate scale. I split the page into two sections in order to tell this story from two perspectives: the tornado's, and the chase team's. Late in our process we obtained a better analysis of the sub-vortices circling around the main tornado, which we added as a detailed inset.

THE TORNADO

6:03:49 P.M.
The twister is unusually wide from the second it touches down—a half mile across. Multiple vortices appear and disappear, confusing chasers as the tornado moves south-southeast.

6:09
Rain wraps around the tornado, hiding it. Accelerating to 40 to 50 miles an hour, the storm turns east, forming subvortices around the main funnel; some last only seconds.

6:15
Just missing El Reno's airport, the tornado pauses, possibly looping in its track, then reorganizes into a single large vortex that rapidly expands and surges eastward.

6:19
As it crosses Highway 81, the twister veers to the northeast, still expanding. Surface winds feeding the storm uproot trees well beyond the base of the funnel.

6:23
The rain parts, revealing the main vortex. A few minutes later, wind speeds approach 300 miles an hour as the tornado's diameter maxes out.

6:27
Steered by the rotating parent storm, the twister makes a loop north of I-40 and heads east. Having spent 40 minutes on the ground, it finally dissipates at 6:43 p.m.

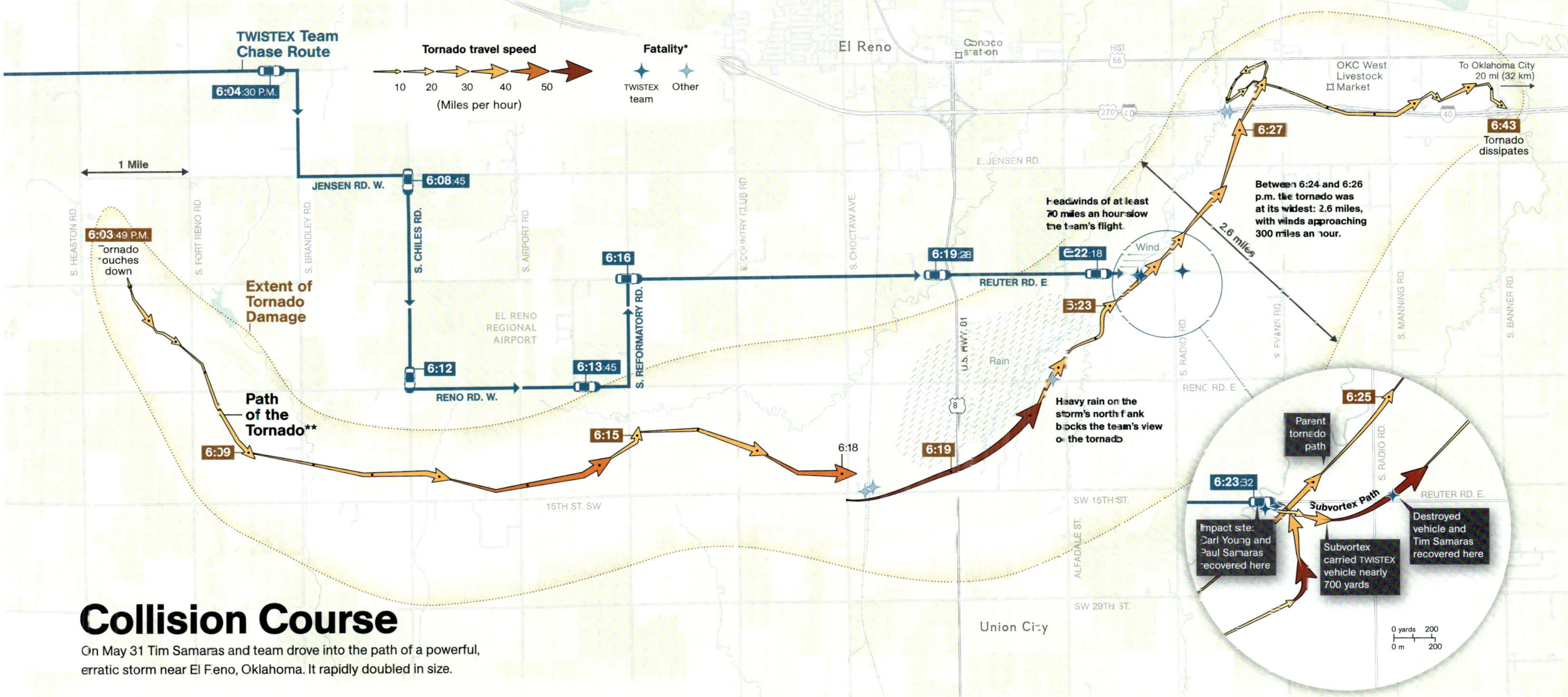

Collision Course

On May 31 Tim Samaras and team drove into the path of a powerful, erratic storm near El Reno, Oklahoma. It rapidly doubled in size.

TIM SAMARAS

6:04:30 P.M.
From the shoulder of a farm road southwest of El Reno, Tim Samaras and his TWISTEX team watch the tornado forming. "This is gonna be a monster wedge," he says.

6:12
Racing to catch up with the fast-moving twister, the team navigates around the airport. But rain obscures their view. "Turn to the north," Samaras tells Carl Young.

6:16
Once on Reuter Road, the team races east, looking for a chance to deploy instrument probes. But they have trouble getting ahead of the tornado on the slippery road.

6:19:28
The team pauses briefly at the Highway 81 intersection, empty of traffic because of closures. The tornado suddenly speeds up. "What a beast," Young says.

6:22:18
Aware of their perilous situation, Samaras urges Young to keep moving. A driver just ahead of them on the road is the last to see them before they disappear in the storm.

6:23:32
As radar imagery later shows, one of several violent subvortices swings around the main vortex and collides with the team's car, flinging it east into a field just past Radio Road.

*The parent storm was responsible for 22 deaths; 8 are shown on this map.

**Generalized tornado track: From start to 6:18 p.m. and 6:26 to end, data collected by the Advanced Radar Research Center of the University of Oklahoma's RaXpol instrument. From 6:18 to 6:26, data provided by the Center for Severe Weather Research's Doppler on Wheels.

MARTIN GAMACHE, NGM STAFF
SOURCES: ERIK FOX; GABE GARFIELD, NATIONAL WEATHER SERVICE AND COOPERATIVE INSTITUTE FOR MESOSCALE METEOROLOGICAL STUDIES; TIM MARSHALL, HAAG GLOBAL; DAN ROBINSON, STORM HIGHWAY; JOSHUA WURMAN, CENTER FOR SEVERE WEATHER RESEARCH

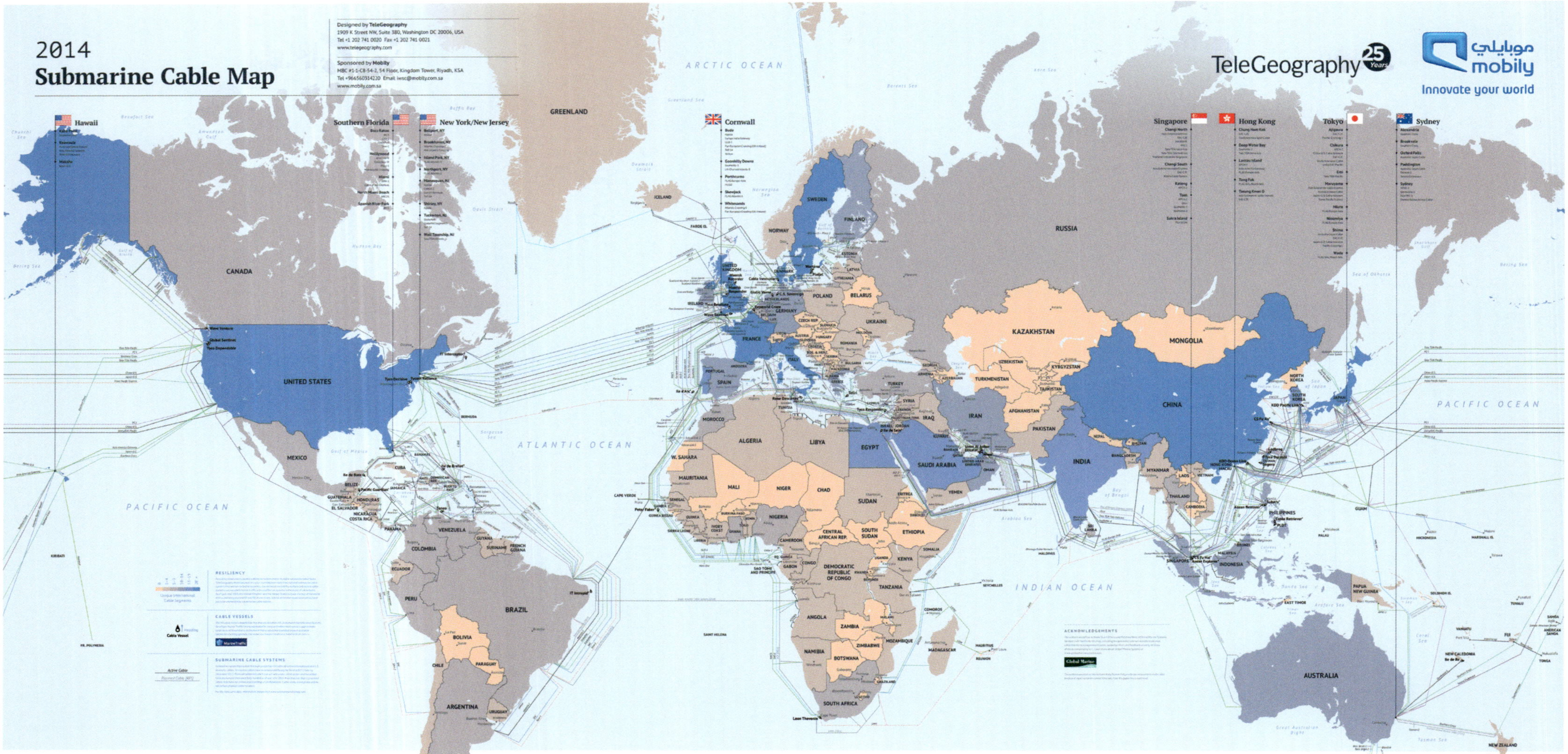

Protectors *of the* Internet

Fiber-optic cables that traverse the bottom of the ocean floor form the backbone of the Internet. This critical global infrastructure relies on a small group of companies responsible for both the installation and maintenance of the more than 300 active submarine cable systems that interconnect the world.

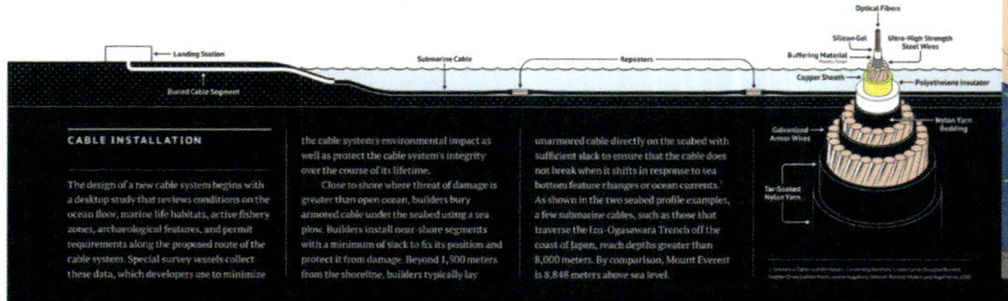

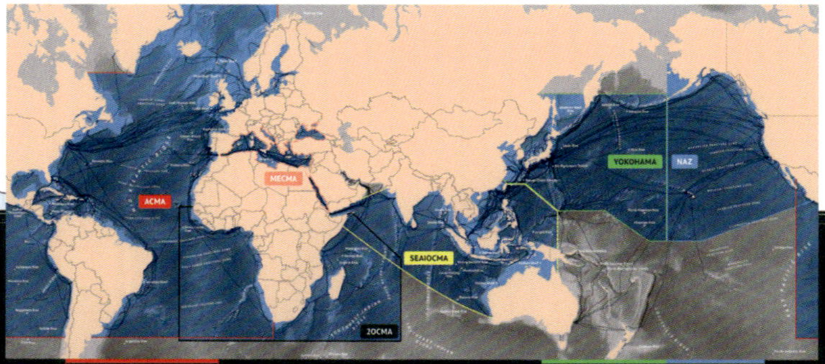

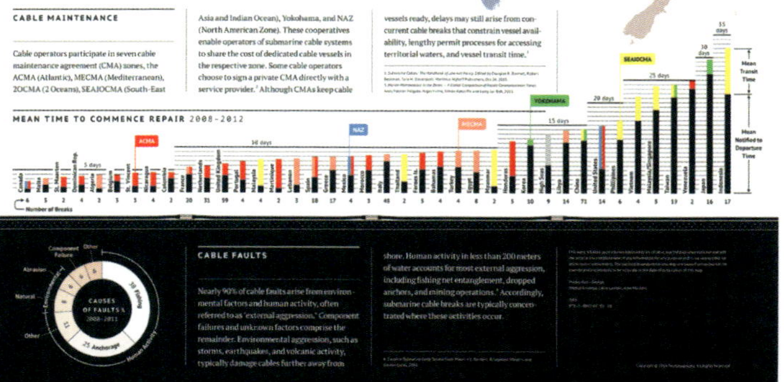

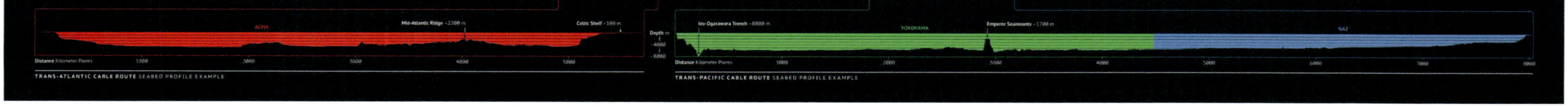

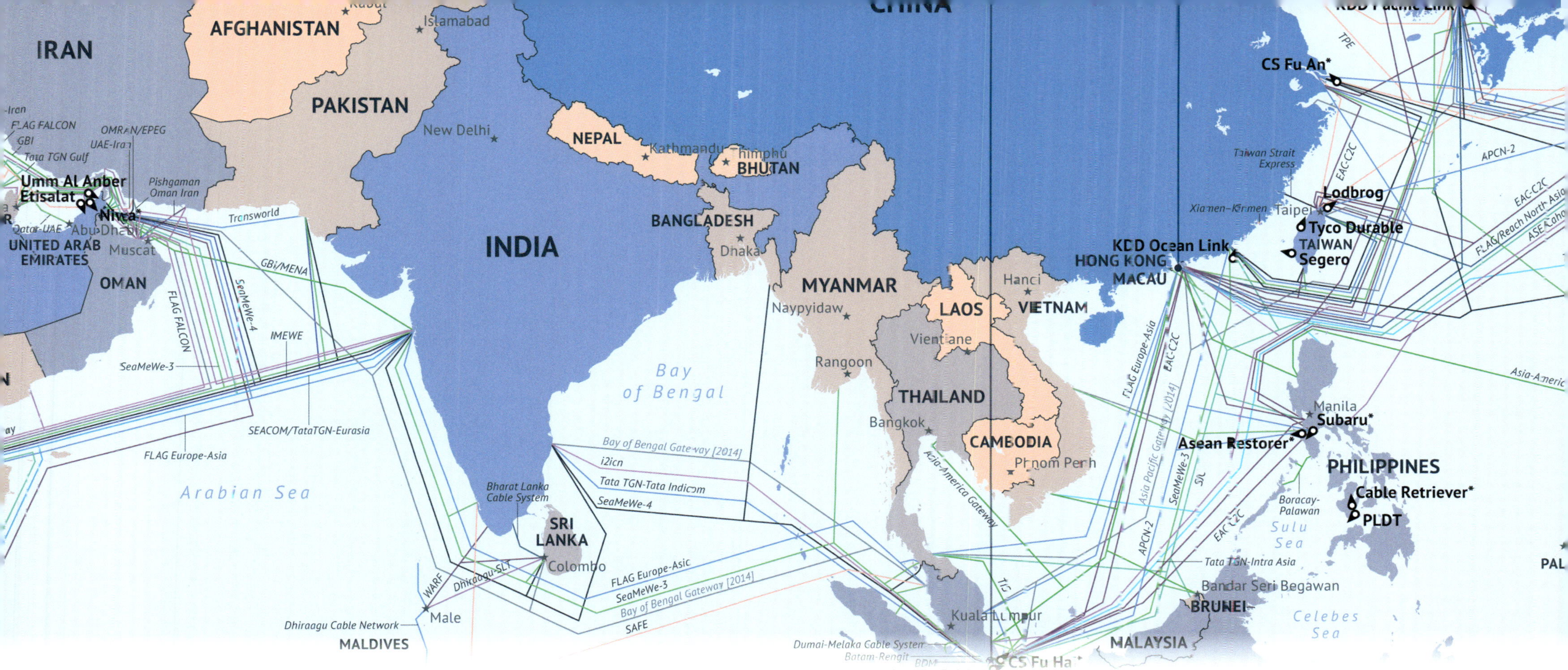

« Submarine Cable Map

TeleGeography

Each year, the telecom market research firm TeleGeography releases its *Submarine Cable Map*, showcasing the physical infrastructure that makes the modern-day Internet possible. The twelfth edition of the map highlights the challenges of installing and maintaining over 300 active cable systems.

The map consists of two sections. The main section shows all active cable systems on a Mercator projection, which was chosen because it provides the most area to draw the numerous overlapping cable systems in Europe and Asia. The secondary section is formatted like a magazine spread with text, illustrations, and infographics. This section explains how a cable system's chosen route takes into account various environmental factors and is designed to minimize damage caused by human activity.

The oceanic and technical theme of the map's content drove the design decisions. The map is dominated by a subtle blue-grey palette with other primary colors used to highlight small details, such as the positions of the cable ships on the main projection. Primary colors in the narrative section are used to represent the same area of the ocean in every instance it's referenced in the seabed profile, inset map, or chart showing average repair times. A legible, modern typeface—PT Sans—was employed throughout the map.

How the map would function as a physical object was also considered. Because of its large dimensions (50 × 36 inches), it was printed on synthetic paper stock, which makes it less prone to creases and tears. In keeping with the theme of its content, the paper is also water resistant.

Tornado Travel »

John **Nelson**

When I look at this map now, I'm encouraged by a desire to do right by a phenomenon and a willingness to start over.

The creation of this map happened at a weird intersection of borrowed urgency and personal reluctance, and is most memorable to me as a bright spot, a turning point that pulled me out of a rut.

Why urgency? Robert Staskowski, a meteorologist with young children, reached out to me the morning after the terrible loss of life in Moore, Oklahoma, when a tornado struck a school, among many other things. His idea was a tornado warning system that was informed by the regional tendencies of tornadoes to follow a certain bearing. Impressed by his motivation and sincerity of vision, I started considering the idea of storms flowing along prevailing directions. This map was an early exploration of that idea.

Why reluctance? The year leading up to this map had been the most difficult but prolific of my life. My parents, both lifelong geography teachers and lovers of maps and travel, were holed up in their northern Michigan home, my father tending to my mother who was living through the final stages of cancer. My feelings of helplessness were pervasive, and I found myself mapping furiously through weekdays, to share on visits; I suppose mostly trying to feel productive in an area I associated with their happiness. Following my mother's death I found that I had fully mined my mapping volition and felt no desire to create. Months went by; I would have ideas but no will. Robert's email and this map's inception mark a moment when I was pulled from a slump and brought back to a place where making felt alright again.

Regarding the process, there are a couple things I recall fondly. In earlier drafts I attempted to interpolate a smooth surface of average tornado direction—a fun idea but ultimately a dead end. Rather, I segregated the individual storms into their general directions as four sibling maps that each wound up communicating regional aspects of movement, all unified by a compass-like chart of prevalence, which doubled as a sneaky legend.

When I look at this map now, I'm encouraged by a desire to do right by a phenomenon and a willingness to start over. But mostly I'm reminded of tentative steps back into a process that I love and the reasons why I love it.

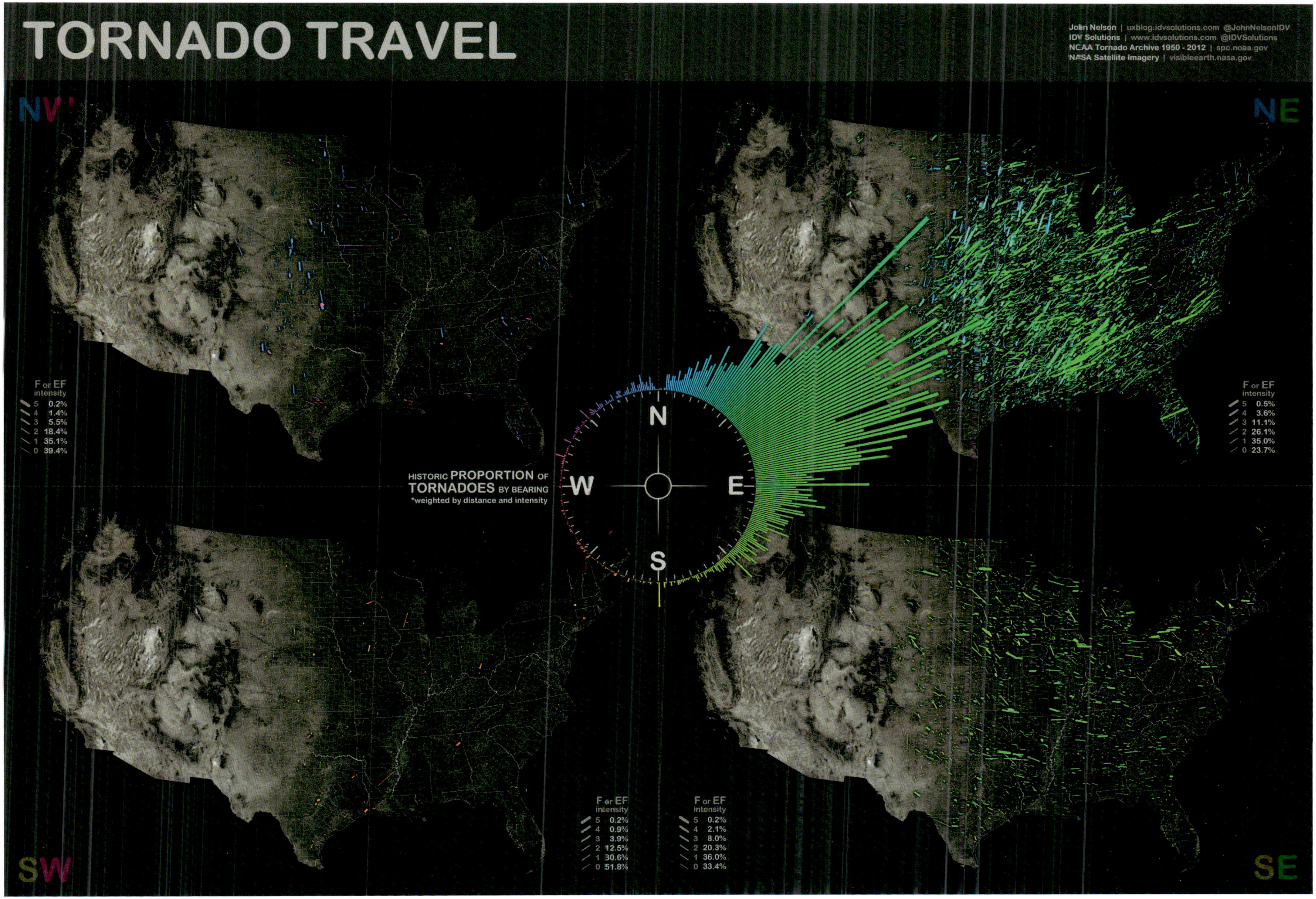
TORNADO TRAVEL
John Nelson | uxblog.idvsolutions.com @JohnNelsonIDV
IDV Solutions | www.idvsolutions.com @IDVSolutions
NCAA Tornado Archive 1950 - 2012 | spc.noaa.gov
NASA Satellite Imagery | visibleearth.nasa.gov
NW
NE
F or EF intensity
5 0.2%
4 1.4%
3 5.5%
2 18.4%
1 35.1%
0 39.4%
F or EF intensity
5 0.5%
4 3.6%
3 11.1%
2 26.1%
1 35.0%
0 23.7%
HISTORIC PROPORTION OF TORNADOES BY BEARING
*weighted by distance and intensity
N
W
E
S
F or EF intensity
5 0.2%
4 0.9%
3 3.9%
2 12.5%
1 30.6%
0 51.8%
F or EF intensity
5 0.2%
4 2.1%
3 8.0%
2 20.3%
1 36.0%
0 33.4%
SW
SE

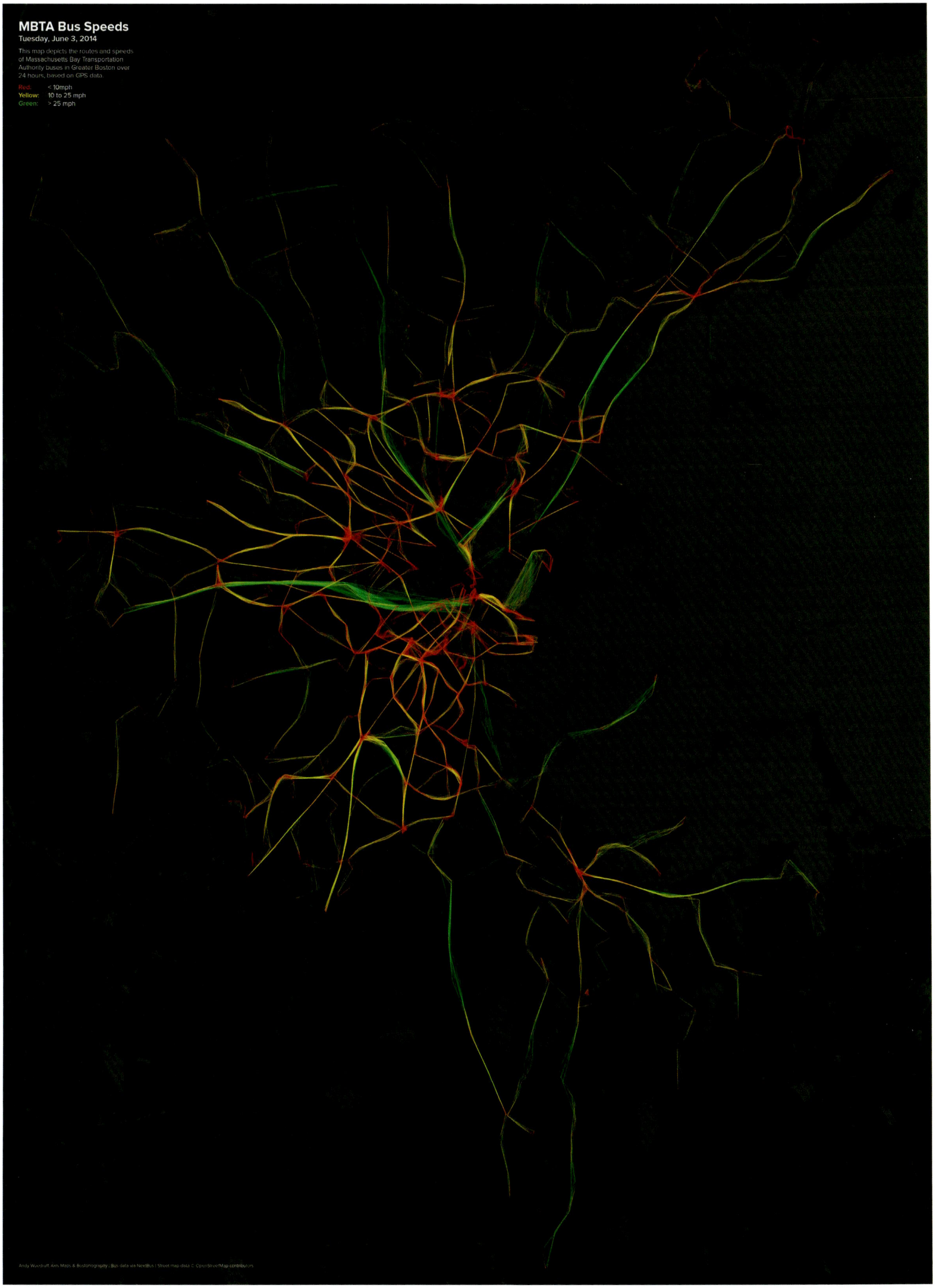
MBTA Bus Speeds
Tuesday, June 3, 2014
This map depicts the routes and speeds of Massachusetts Bay Transportation Authority buses in Greater Boston over 24 hours, based on GPS data.
Red: < 10mph
Yellow: 10 to 25 mph
Green: > 25 mph
Andy Woodruff, Axis Maps & Bostonography | Bus data via NextBus | Street map data © OpenStreetMap contributors

« MBTA Bus Speeds

Andy ***Woodruff***

Exploratory mapping is an excellent way to learn about a city, and that purpose—to understand more about my own city—is what led to this map. Entirely inspired by some of Eric Fischer's work with San Francisco data, it shows a cumulative view of twenty-four hours of bus service in and around Boston. It's one of an ongoing map series of hourly and daily movements by Massachusetts Bay Transportation Authority (MBTA) buses. With publicly available real-time location data, it is possible to create a portrait not only of where buses go, but also how frequently they go to those places and how fast they move. Seen here are several hundred thousand reported bus location points, strung together into route sequences and drawn as nearly transparent lines.

A simple plot of the data has an inherent, organic beauty in its forms. Routes follow the twisted networks of a city with decidedly non-orthogonal road geometry. Wispy tendrils reach into far-flung areas with little bus service, where opacity on the map does not accumulate. Web-like structures appear where GPS signals were imprecise or lost, often underground or underwater. It all seems appropriate for a chaotic, living city.

Bright traffic light colors are meant to invoke a familiar speed metaphor, and also simply to be visually striking. A certain degree of pure "eye candy" in a map is a useful thing. Here my hope is that the map reader not only will understand a little more about the geography of public transportation in Boston, but also will be captivated by the aesthetics and inspired to go learn even more—to ask *"why?"*

A certain degree of pure "eye candy" in a map is a useful thing.

» *Since the geographic information is very simple, the map must have a lower voice.*

O Mapa do Tráfico (The Traffic Map) »

Eduardo **Asta**

This was a very interesting project because, while the data need to be geographically located, they are so simple that a very realistic or detailed map is not necessary. Even a map is not necessary: we could place all this information on a diagram, for example.

In this case, why should we use a map? Because it is a universal graphical interface, well known by most audiences. Having a map helps you to understand that something is happening across the world, even if you do not speak the language used in the infographic. The map behind these traffic routes tells you that something is starting in this area and ending in these other ones.

Since the geographic information is very simple, the map must have a lower voice. I selected only the information I most needed for the map, and I built the first layer of my graphic using black and grey fills and black strokes.

The second layer has the main information: routes of cocaine and heroin—their origins, paths, and destinations. Here we have all the joy of this infographic. Because it very colorful and bright while the map is very dark, the second layer jumps forward at the audience. It is seen before anything else. I see this as a powerful feature since the main information can be read fast. The reader is grateful to have what they want without obstacles, and I can say the reader will have a good experience reading it.

The third layer introduces producers of heroin and cocaine, the market share of four kinds of drugs, and some supportive text.

I think that this map applies simple and useful reading principles. It was made to be read by regular people, the ones that are not trained in graphic design. I like to think that my mom could read it and explain it to her friends.

DROGAS

O MAPA DO TRÁFICO

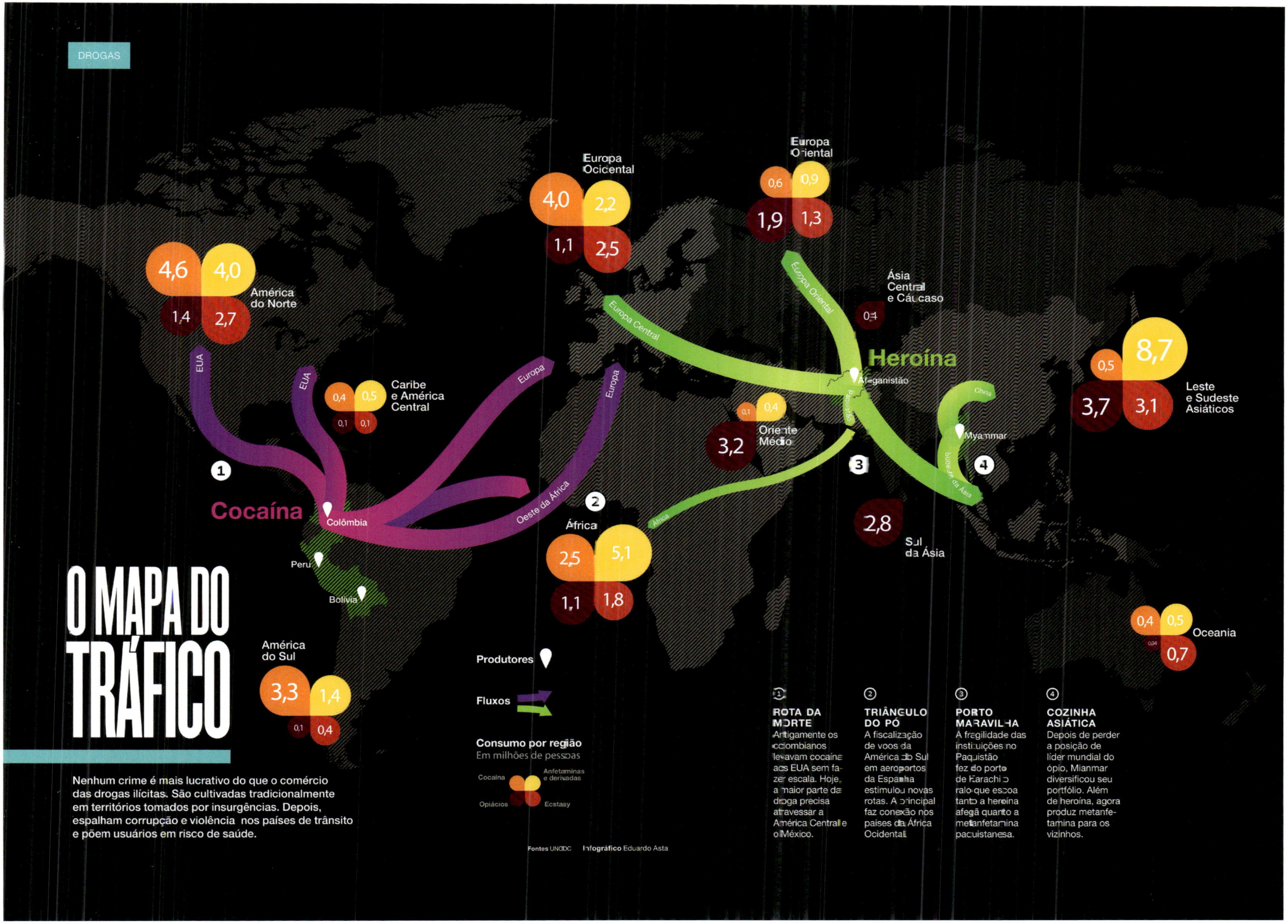

Produtores

Fluxos

Consumo por região
Em milhões de pessoas

Cocaína · Anfetaminas e derivadas · Opiácios · Ecstasy

Nenhum crime é mais lucrativo do que o comércio das drogas ilícitas. São cultivadas tradicionalmente em territórios tomados por insurgências. Depois, espalham corrupção e violência nos países de trânsito e põem usuários em risco de saúde.

Fontes UNODC   Infográfico Eduardo Asta

① **ROTA DA MORTE**
Antigamente os colombianos levavam cocaína aos EUA sem fazer escala. Hoje, a maior parte da droga precisa atravessar a América Central e o México.

② **TRIÂNGULO DO PÓ**
A fiscalização de voos da América do Sul em aeroportos da Espanha estimulou novas rotas. A principal faz conexão nos países da África Ocidental.

③ **PORTO MARAVILHA**
A fragilidade das instituições no Paquistão fez do porto de Karachi o ralo que escoa tanto a heroína afegã quanto a metanfetamina paquistanesa.

④ **COZINHA ASIÁTICA**
Depois de perder a posição de líder mundial do ópio, Mianmar diversificou seu portfólio. Além de heroína, agora produz metanfetamina para os vizinhos.

European subterranean veins

Have you ever considered travelling from London to Paris by Tube? How long are the underground "veins" that run below major European cities? This analysis and visualization compares the distances covered and journey fares of various underground systems across the continent, revealing some interesting insights and comparing actual sizes on the map.

How to read it?

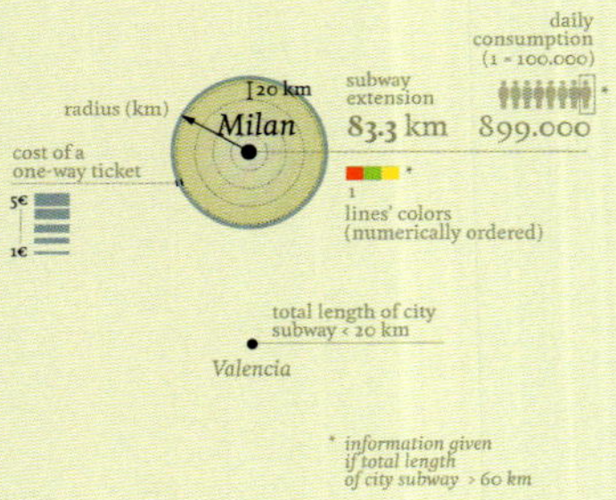

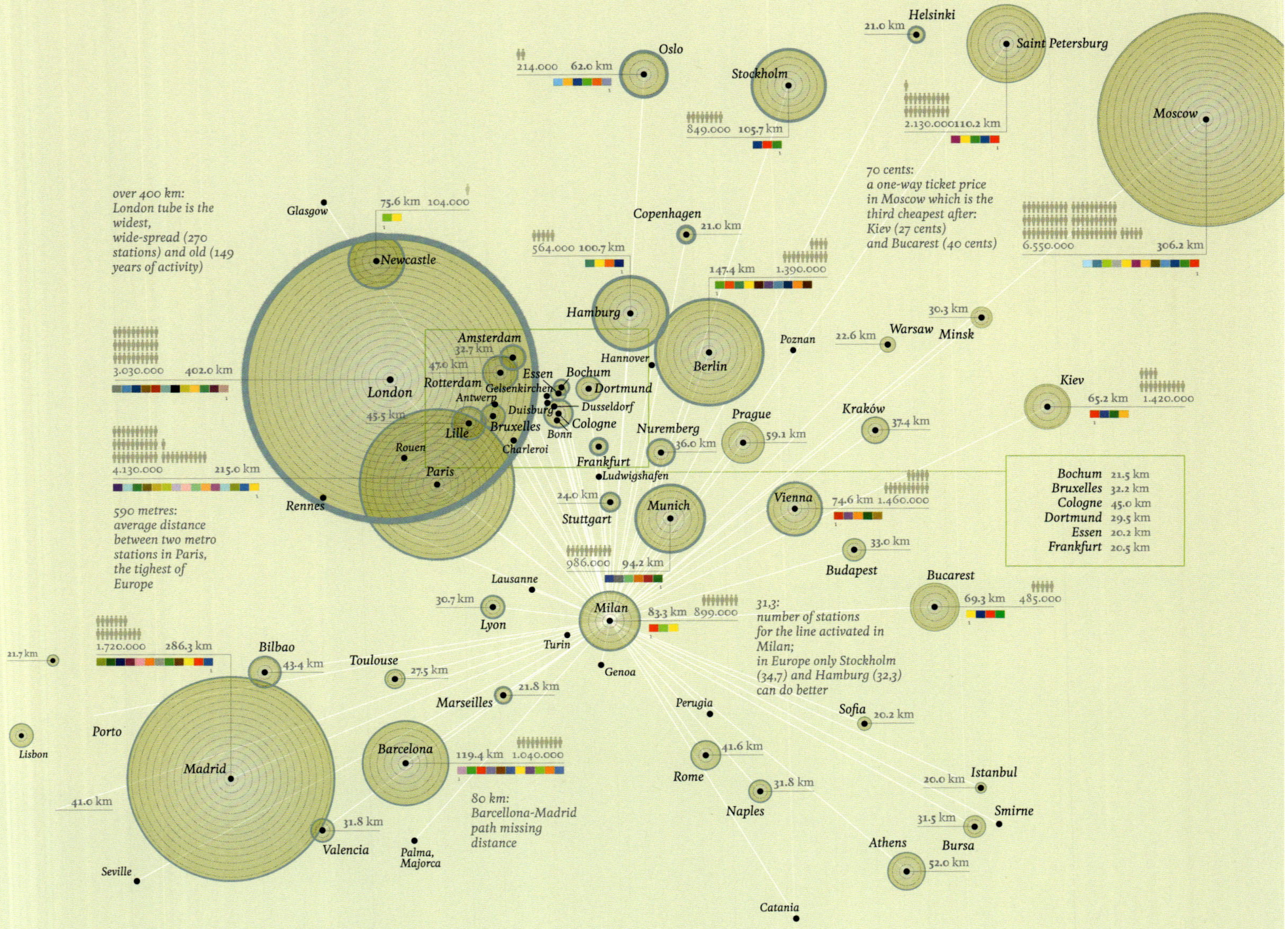

Subway length compared with...

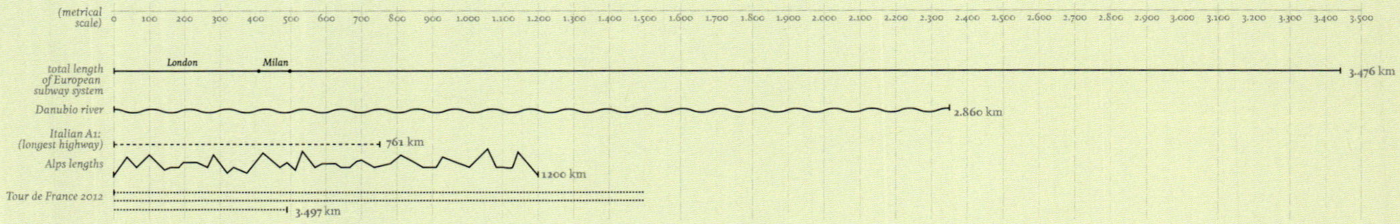

The visualization has been designed and produced by Accurat (www.accurat.it), and was originally published in italian on La Lettura the sunday cultural supplement of Corriere della Sera.

« European Subterranean Veins

Accurat

Have you ever considered travelling from London to Paris by Tube? How long are the underground "veins" that run below major European cities? This analysis and visualization compares the distances covered and journey fares of various underground systems across the continent, revealing some interesting insights and composing an imaginary European travel map where we cut the morphological layer.

Each city is represented by a circle starting from its centre, the dimension of which represents the sum of its subway extension. Isn't it compelling that most European cities could be reached and can overlap with the subway?

As for the general layout, we chose to cut the geographical layer of the visualization, in order to highlight this imaginary idea of travelling below the earth, and we simply positioned cities on the map without the morphology. A brief idea of each city distance from Milan (the headquarters of our publisher) is illustrated through light white lines in the background.

The visualization then provides lots of other information: each city subway system can be further compared, and readers could spend additional time looking at daily usage, the actual colors of the lines, and the cost of a one-way ticket. On the bottom right of the page we also added a visual comparison to give people an idea of actual lengths. We compared the total sum of all the European subways with some well-known and easily pictured linear distances using the same scale of the main visualisation. Did you know that the total length of the European subway veins is way longer than the Danube River, or the main Italian highway? And how about the lengths of the Tour de France or the Alps? European subways win again.

Sometimes it happens that ideas come as an enlightenment, and the overall "architecture" of the visualisation is driven by sudden intuitions you might have without even knowing whether you'll be able to find the data. We were reading that London Underground's total length is approximately 400 km. Isn't it huge? How far can you go in 400 km? What if it was straight? Could we get over the English Channel? Could we reach Paris? Yes, the visualization was almost there already.

Sometimes it happens that ideas come as an enlightenment... »

Manhattan 311 Noise Complaints »

Karl **Sluis**

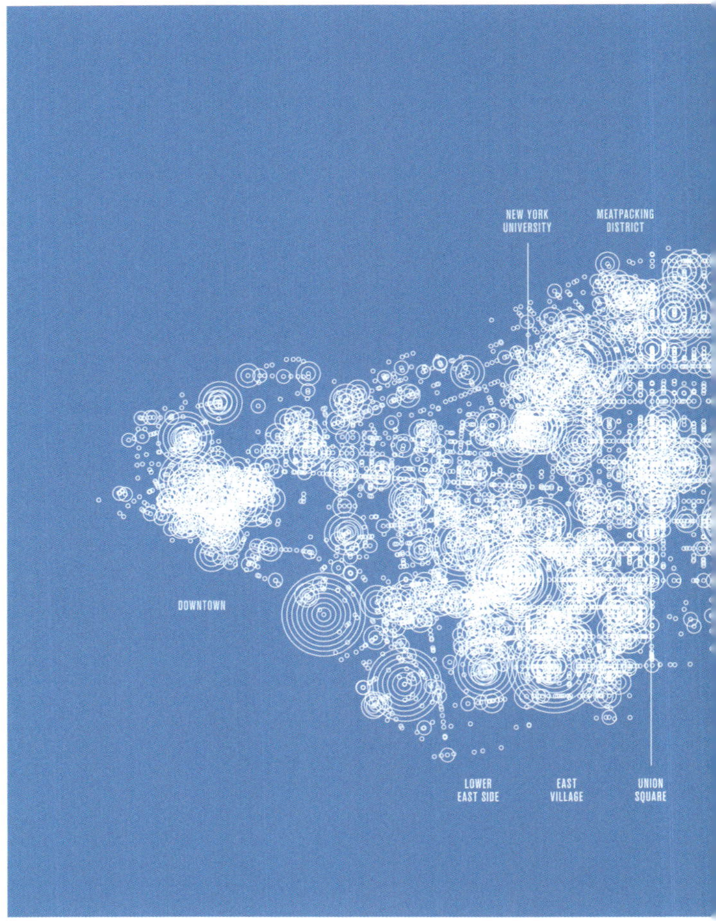

Often, great design depends on the simplicity and clarity that can only be achieved through elimination. "Perfection is achieved, not when there is nothing more to add, but when there is nothing left to take away," writes Antoine de Saint-Exupéry in *Airman's Odyssey*. This map of Manhattan reveals the structure of the city through one dimension: noise complaints to New York City's information hotline, 311.

This map is notable, and beautiful, as much for what it lacks as for what it marks. You'll find no streets, no subways, no parks, trees, compass, elevations, or grid—just a handful of neighborhoods annotated to help orient viewers. It relies on the iconographic shape of Manhattan and the metaphor of sound waves traveling through the air, like ripples on the surface of a pond. Intuitively, a larger group of circles indicates more noise complaints at that location on the map, and more interference between circles marks a noisier neighborhood. The overlaps of these circles and lines create unique patterns and textures that draw in viewers.

Mapped out, noise complaints alone can expose an incredible number of overlapping structures, from the rigid gridiron of the city's streets to where people live, work, and play—or even all three. The noisiest neighborhoods (the Lower East Side, Times Square) are those where entertainment and housing collide, often above one another. By contrast, Midtown, the epicenter of work, and the Upper East and West Sides, highly residential neighborhoods both, barely register on the map. Noise reveals socioeconomic priorities of the city, as construction complaints increase south of Central Park. Even the ongoing development of the Second Avenue subway line appears on the map. It's the rhythm of noise in the city, the balance of quiet and cacophony, that makes living in New York City, and this map, so unique and compelling.

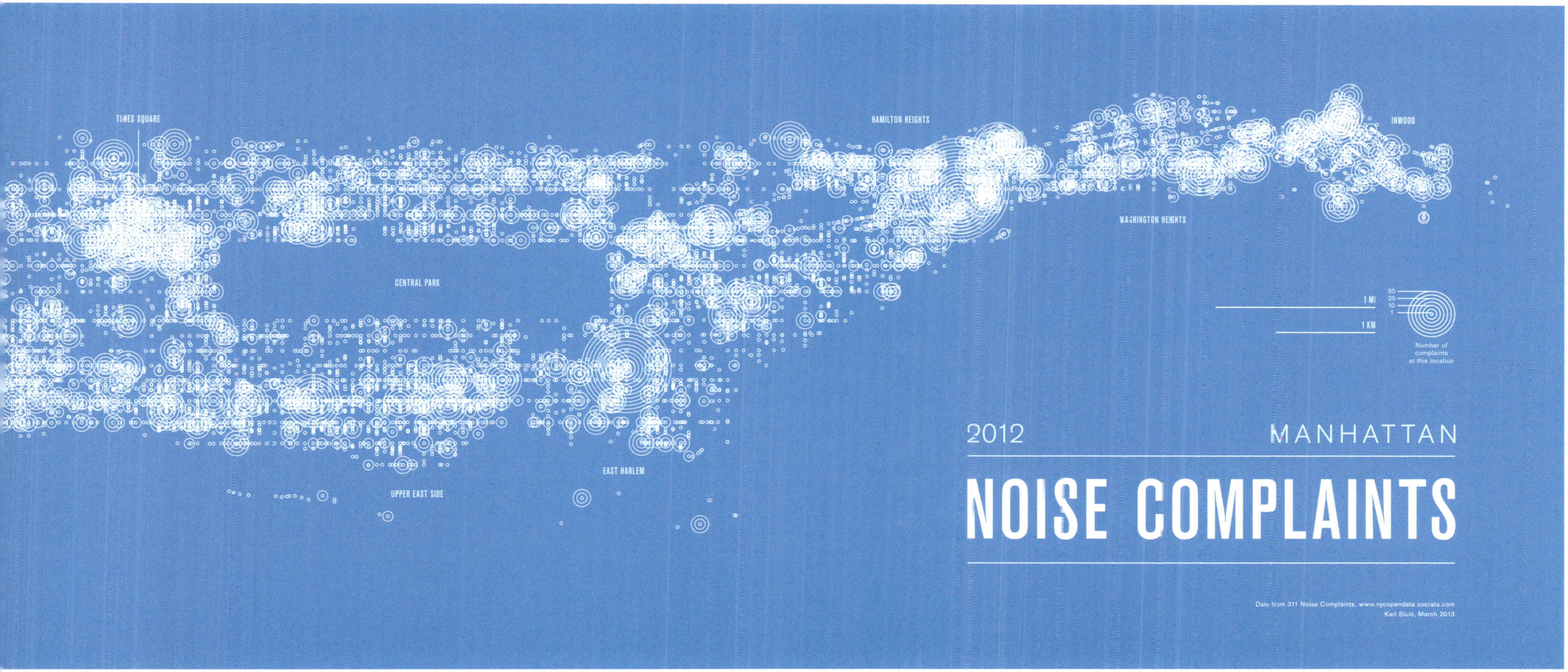
TIMES SQUARE
HAMILTON HEIGHTS
INWOOD
WASHINGTON HEIGHTS
CENTRAL PARK
EAST HARLEM
UPPER EAST SIDE
1 MI
1 KM
50
25
10
1
Number of complaints at this location
2012
MANHATTAN
NOISE COMPLAINTS
Data from 311 Noise Complaints, www.nycopendata.socrata.com
Karl Sluis, March 2013

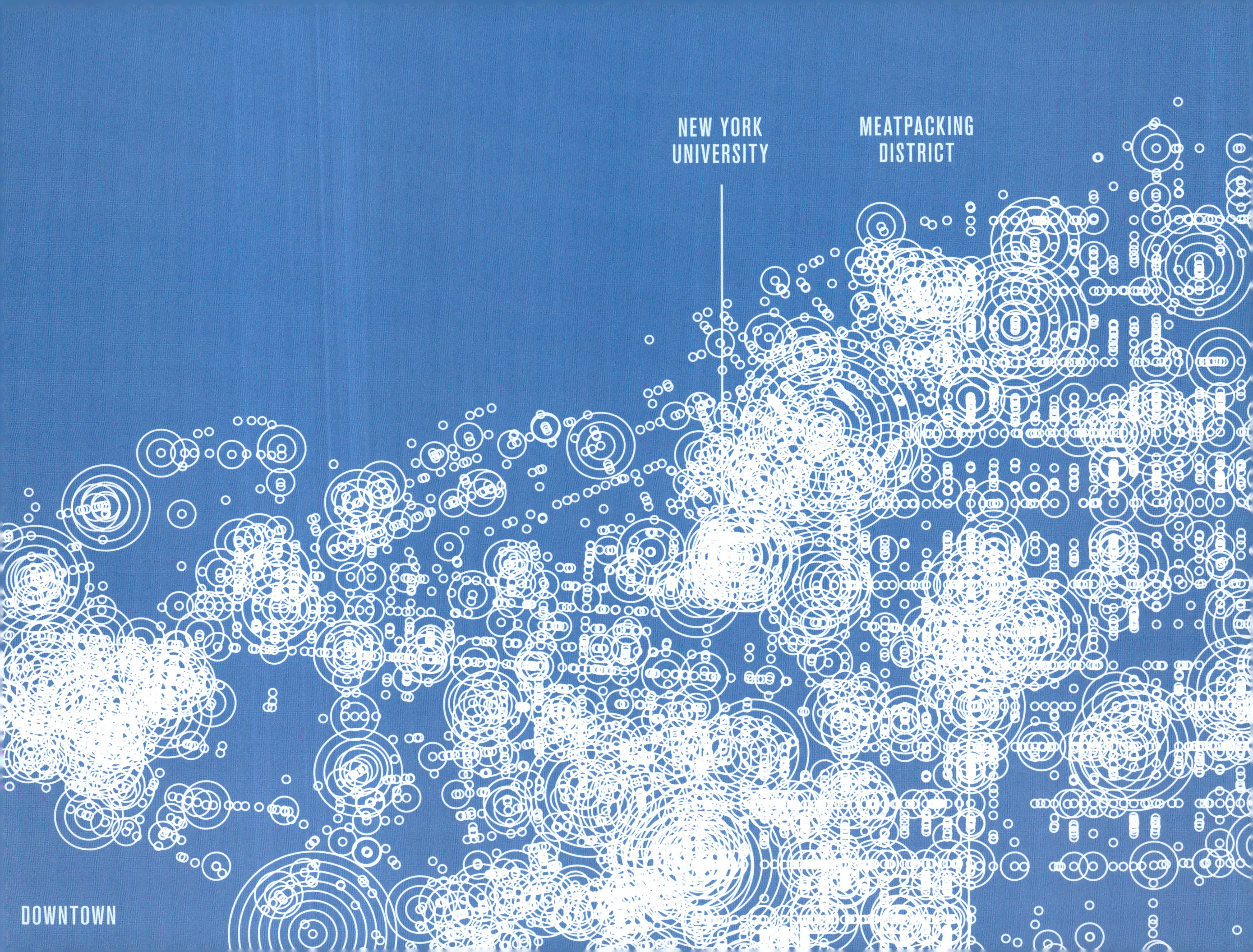
NEW YORK
UNIVERSITY
MEATPACKING
DISTRICT
DOWNTOWN

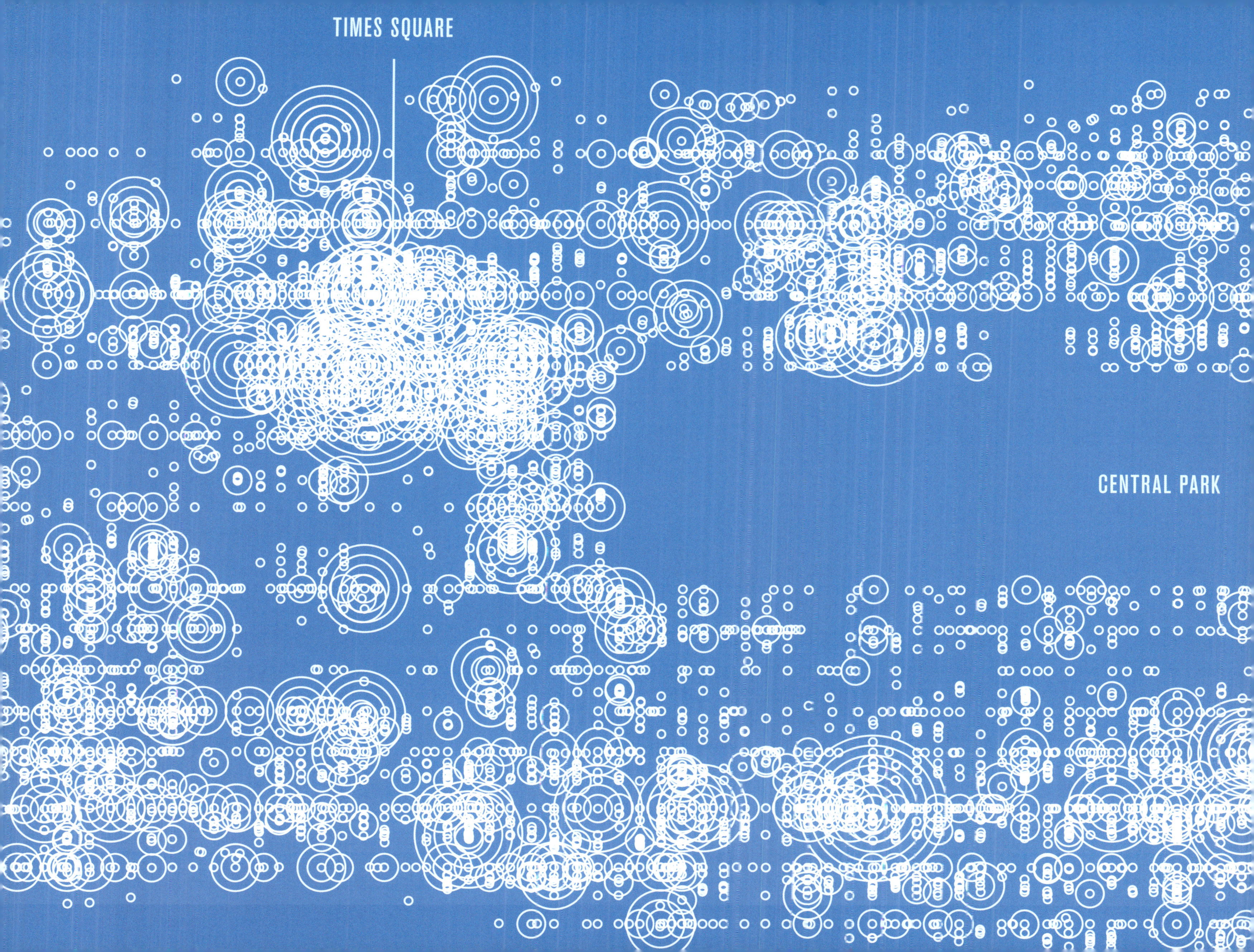
TIMES SQUARE
CENTRAL PARK

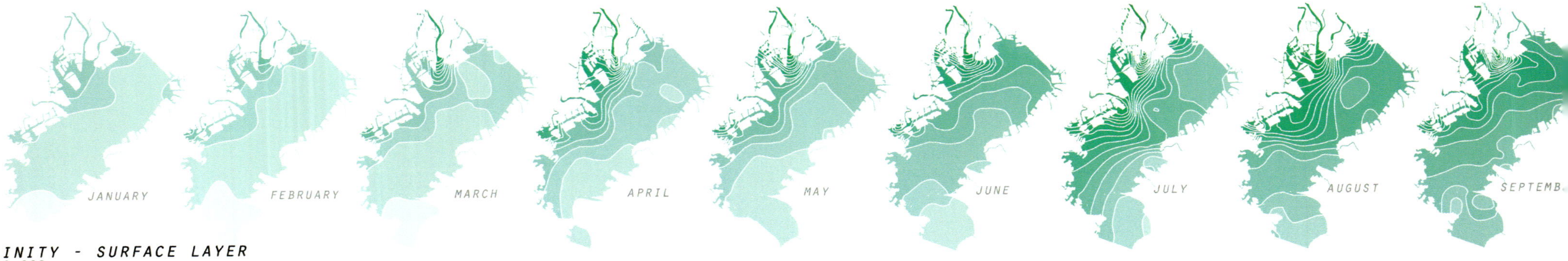

SALINITY - SURFACE LAYER
1:600,000

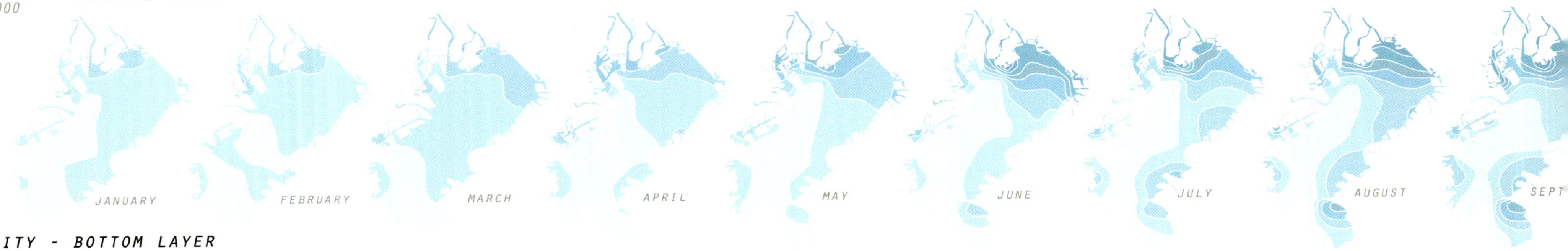

SALINITY - BOTTOM LAYER
1:600,000

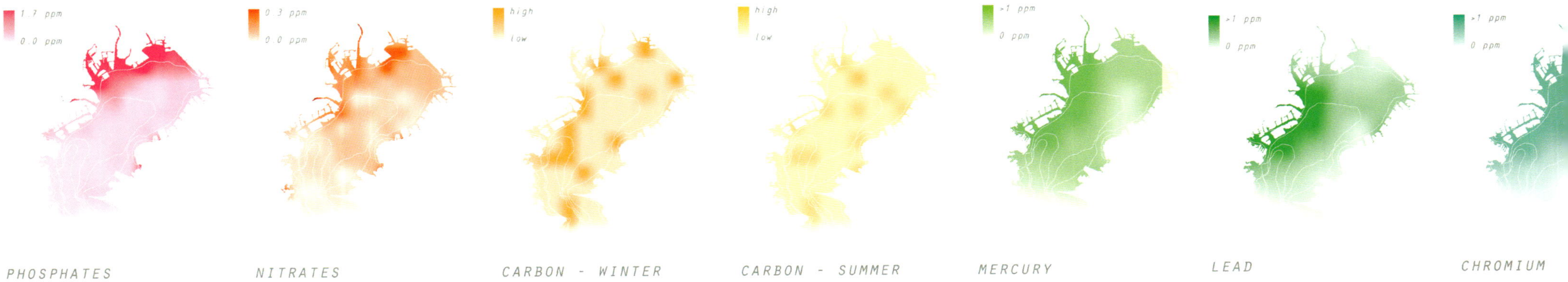

WATER QUALITY - POLLUTION
1:600,000

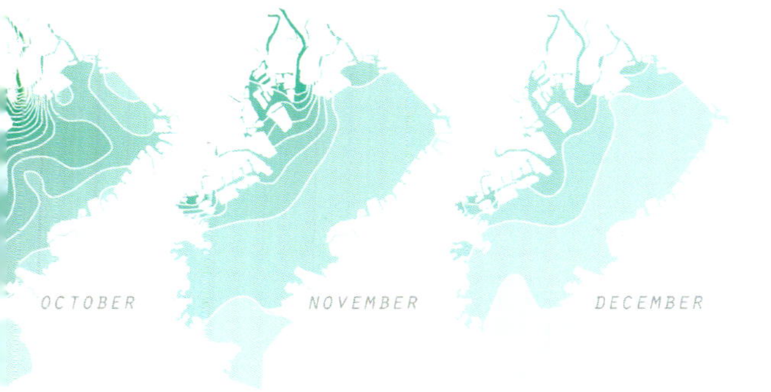

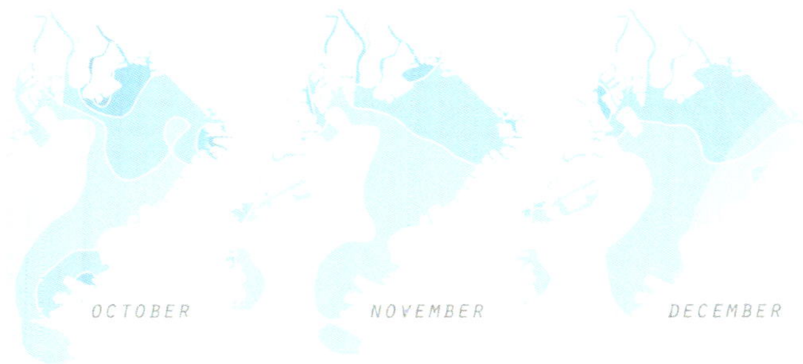

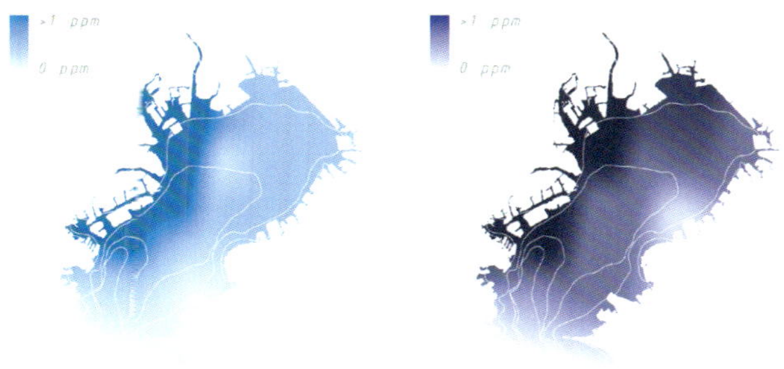

CADMIUM

ARSENIC

« The Hydrological Gradients of Tokyo Bay

Andrea **Hansen**

The Tokyo Bay region is the largest and most populous industrial region in Japan, with over 35 million people living in the Greater Tokyo metropolitan area. As a result of such intense industrialization and urbanization, the bay has experienced a remarkable transformation. What was once a soft-edged coastline is now fortified with megastructures and land reclaimed from the sea. Meanwhile, significant air and water pollution have caused a near-complete cessation of fishing in the bay, thus forcing the Japanese fishing industry much farther out to sea.

Mapping conditions in such a dynamic environment required the application of unconventional mapping techniques that address the ineffectiveness of traditional notation at depicting fluid systems. Standard hydrological measuring conventions such as isobaths, isobathytherms, or isohalines—all variations of the contour line—are too rigid to express the constant flux of a body of water. In contrast, this set of diagrams renders hydrological data as soft yet precise gradients, using Tufte's principle of small multiples to reveal small changes in the concentration of salinity, sediment, or industrial pollutants over time.

My design was influenced by Jacques Bertin's retinal variables—singular visual elements such as size, shape, or color that can be varied to express quantitative change. Keeping in mind that these maps are meant to be read as a series rather than as individuals, only one retinal variable at a time was used to show change across the series. Color is the variable in the water quality series, since each map shows a different type of pollutant. In the salinity maps, meanwhile, I adjusted only value (brightness), since at this scale the eye can discern value changes more easily than color or pattern changes. I also avoided labels, since showing relative (seasonal) variation was more important than highlighting the salinity at any one point in place or time.

Mapping information with gradients makes sense for a hydrological body, in which exact measurements are impractical given the fluidity of the system. It also reveals important findings about the greater hydrology of Tokyo Bay: as expected, pollution is higher along urban edges, but these areas also have a number of river deltas where there is an influx of freshwater in the spring and summer, and an upwelling of subsurface saltwater in autumn as the colder freshwater at the surface layer sinks.

A Typographic Map of the World »

Andrew E. **Burgess**

This map of the world was created with the intention of balancing visual aesthetics and conveyance of information, while limiting the medium to typography. Its inherent beauty is in its elegance; it is merely a collection of names, yet with careful attention given to proportion and adjacency. The black and white quality heightens its purity and simplicity.

The map changes behavior at various scales. Viewed from very close, information about population density and geographical adjacency is discernible. As the viewer's perspective widens, the individual characters are less prominent and the text begins to form shapes familiar to most as landmasses and continents. This approach is akin to pointillism.

Three different typefaces were used: one for country names, one for city names (italicized for capital cities), and one for bodies of water. These typefaces were carefully considered in order to convey a sense of hierarchy—the names of countries, for example, are capitalized and bolder than the font chosen for names of cities. Additionally, larger nations in terms of land area employ a slightly greater tracking. Major rivers are expressed in loose representations of their channels, and oceans are labeled in sweeping curves for visual effect. Latitude and longitude designations appear around the border, providing an illusory frame.

Creating a map in this medium presented some unique challenges. Sizing of letters posed one such difficulty—the typefaces had to be small enough to convey contours of shorelines while retaining readability. This led to consideration of the overall dimensions of the map and the scale at which the continents were to be represented. Another difficulty resided in the placement of text in accordance with actual geographic location; some cities have long names or are close in proximity to other cities. In more densely populated areas city locations are approximate, illustrating both the relativity of the medium as well as the interplay between data and aesthetics.

The final result seeks to conceal the subtleties incorporated in creating this map. Yet, though they are barely perceptible, these details provide the foundation for its elegance.

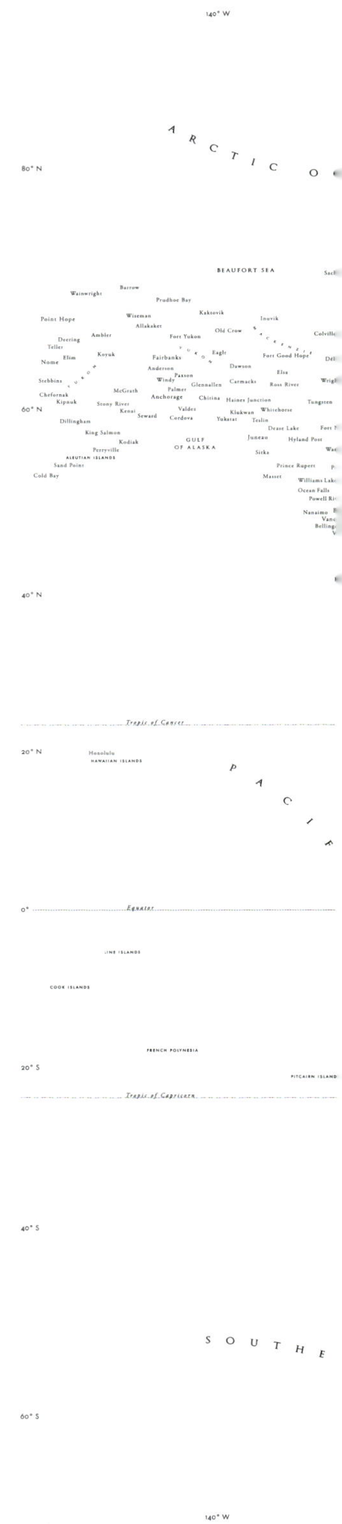

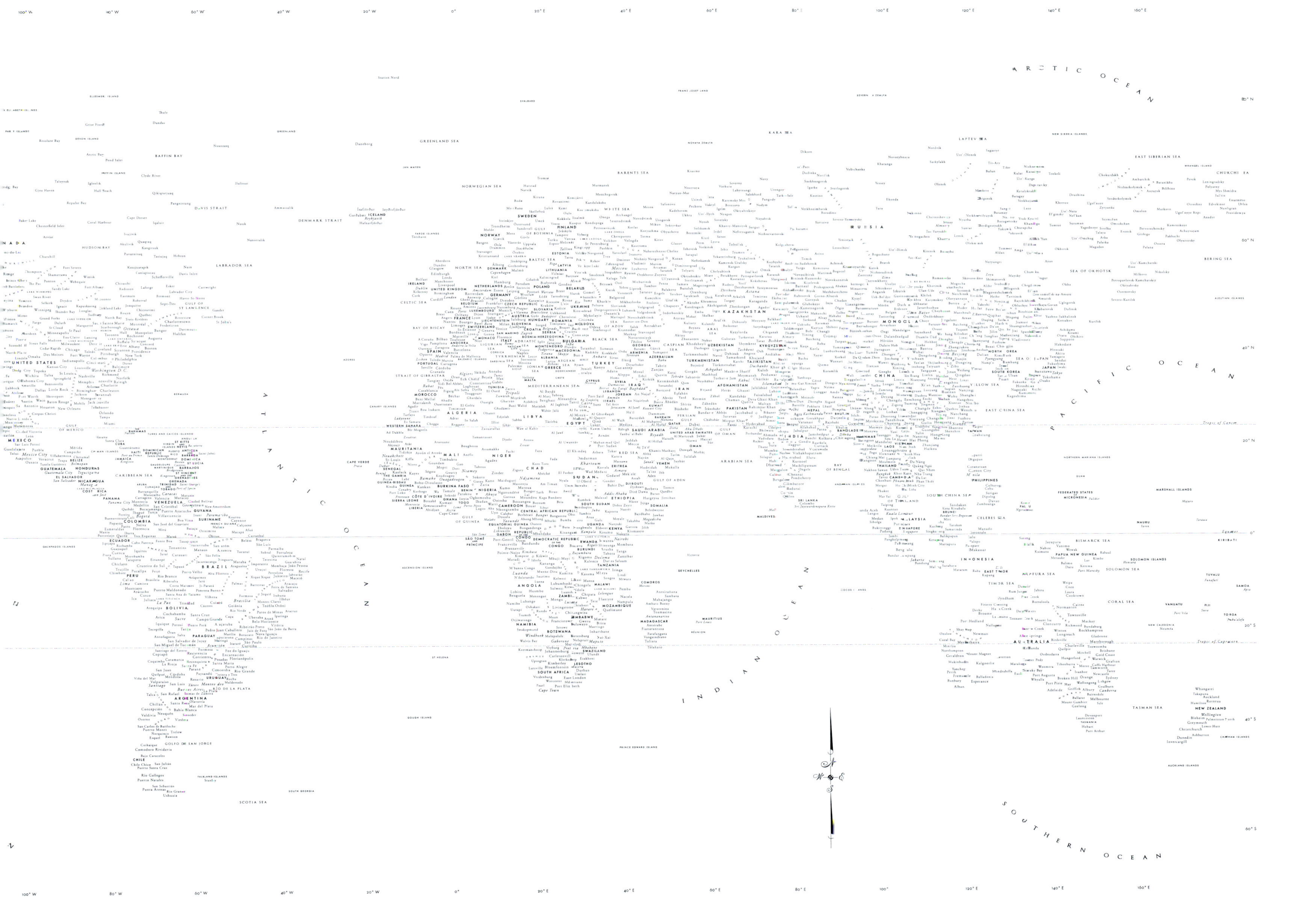

100° W
80° W
60° W
40° W
20° W
0°
20° E
40° E
60° E
80° E
100° E
120° E
140° E
160° E
80° N
60° N
40° N
20° N
0°
20° S
40° S
60° S
ARCTIC OCEAN
PACIFIC OCEAN
ATLANTIC OCEAN
INDIAN OCEAN
SOUTHERN OCEAN
GREENLAND SEA
NORWEGIAN SEA
BARENTS SEA
KARA SEA
LAPTEV SEA
EAST SIBERIAN SEA
BAFFIN BAY
DAVIS STRAIT
DENMARK STRAIT
LABRADOR SEA
HUDSON BAY
NORTH SEA
BALTIC SEA
BAY OF BISCAY
MEDITERRANEAN SEA
BLACK SEA
CASPIAN SEA
RED SEA
ARABIAN SEA
BAY OF BENGAL
SOUTH CHINA SEA
EAST CHINA SEA
SEA OF OKHOTSK
BERING SEA
CARIBBEAN SEA
CORAL SEA
TASMAN SEA
SCOTIA SEA
Tropic of Cancer
Tropic of Capricorn
Equator
ICELAND
NORWAY
SWEDEN
FINLAND
RUSSIA
KAZAKHSTAN
MONGOLIA
CHINA
JAPAN
INDIA
PAKISTAN
AFGHANISTAN
IRAN
IRAQ
SAUDI ARABIA
TURKEY
EGYPT
LIBYA
ALGERIA
MALI
NIGER
CHAD
SUDAN
ETHIOPIA
KENYA
TANZANIA
ANGOLA
NAMIBIA
BOTSWANA
SOUTH AFRICA
MADAGASCAR
MOZAMBIQUE
SPAIN
FRANCE
GERMANY
POLAND
UKRAINE
UNITED STATES
MEXICO
CUBA
COLOMBIA
VENEZUELA
BRAZIL
PERU
BOLIVIA
PARAGUAY
ARGENTINA
CHILE
URUGUAY
INDONESIA
AUSTRALIA
NEW ZEALAND
PAPUA NEW GUINEA
PHILIPPINES

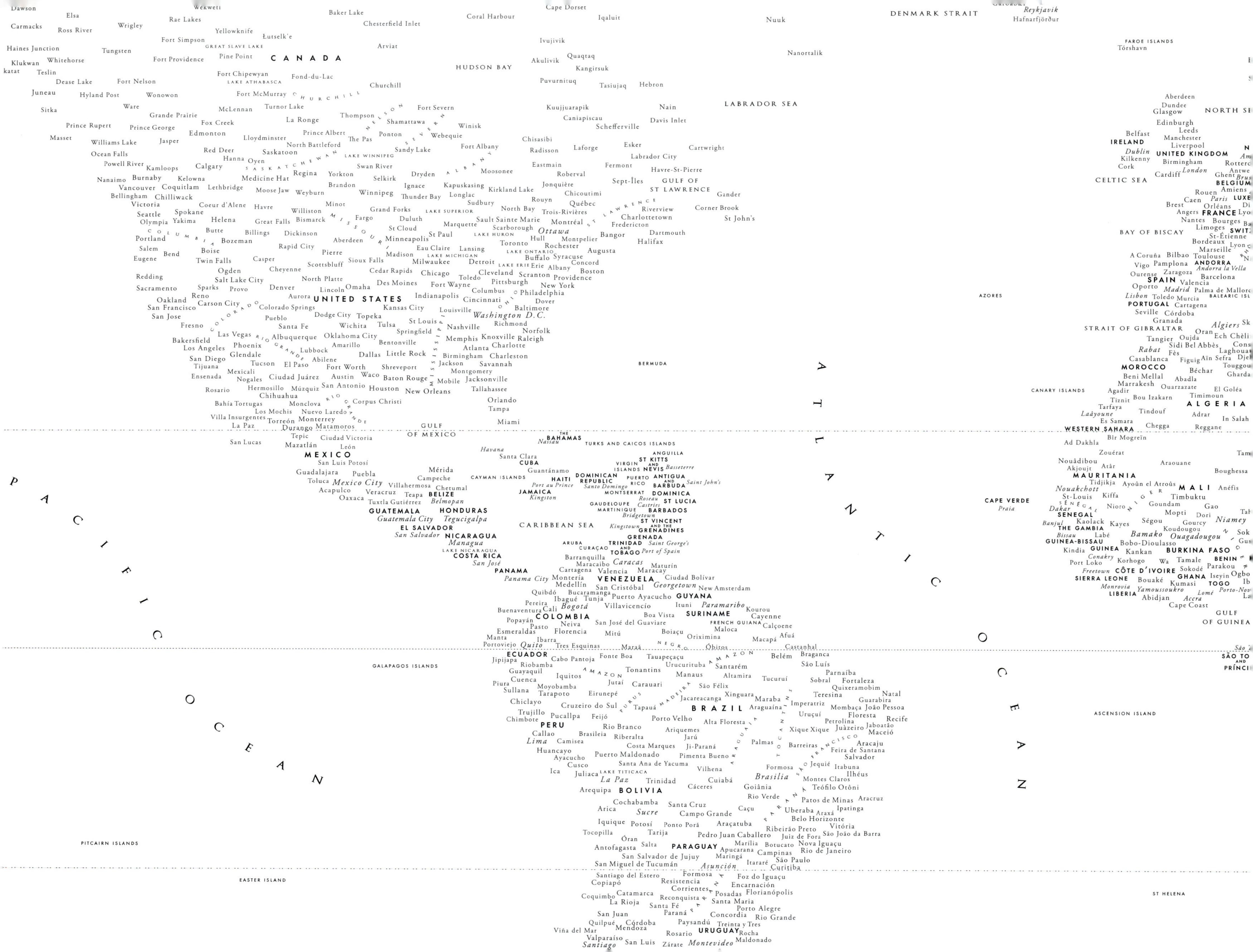

CANADA
HUDSON BAY
LABRADOR SEA
DENMARK STRAIT
NORTH SEA
CELTIC SEA
BAY OF BISCAY
GREAT SLAVE LAKE
LAKE ATHABASCA
LAKE WINNIPEG
LAKE SUPERIOR
LAKE HURON
LAKE MICHIGAN
LAKE ONTARIO
LAKE ERIE
GULF OF ST LAWRENCE
UNITED STATES
GULF OF MEXICO
MEXICO
CARIBBEAN SEA
PACIFIC OCEAN
ATLANTIC OCEAN
BERMUDA
AZORES
CANARY ISLANDS
CAPE VERDE
Praia
GALAPAGOS ISLANDS
PITCAIRN ISLANDS
EASTER ISLAND
ASCENSION ISLAND
ST HELENA
FAROE ISLANDS
Tórshavn
STRAIT OF GIBRALTAR
GULF OF GUINEA
SÃO TOMÉ AND PRÍNCIPE
Dawson
Elsa
Carmacks
Ross River
Wrigley
Rae Lakes
Wekweti
Baker Lake
Chesterfield Inlet
Coral Harbour
Cape Dorset
Iqaluit
Nuuk
Reykjavík
Hafnarfjörður
Yellowknife
Łutselk'e
Haines Junction
Tungsten
Fort Simpson
Arviat
Ivujivik
Nanortalik
Klukwan
Whitehorse
Fort Providence
Pine Point
Akulivik
Quaqtaq
Kangirsuk
Teslin
Dease Lake
Fort Nelson
Fort Chipewyan
Fond-du-Lac
Churchill
Puvurnituq
Tasiujaq
Hebron
Juneau
Hyland Post
Wonowon
Fort McMurray
CHURCHILL
Sitka
Ware
McLennan
Turnor Lake
Fort Severn
Kuujjuarapik
Nain
Grande Prairie
Thompson
NELSON
Caniapiscau
Davis Inlet
Prince Rupert
Prince George
Fox Creek
La Ronge
Shamattawa
Winisk
Schefferville
Masset
Williams Lake
Jasper
Edmonton
Lloydminster
Prince Albert
The Pas
Ponton
SEVERN
Webequie
Chisasibi
Ocean Falls
Red Deer
Saskatoon
North Battleford
Sandy Lake
Fort Albany
Radisson
Laforge
Esker
Cartwright
Hanna
Oyen
SASKATCHEWAN
Labrador City
Powell River
Kamloops
Calgary
Swan River
ALBANY
Moosonee
Eastmain
Fermont
Havre-Saint-Pierre
Nanaimo
Burnaby
Kelowna
Medicine Hat
Regina
Yorkton
Selkirk
Dryden
Roberval
Vancouver
Coquitlam
Lethbridge
Moose Jaw
Weyburn
Brandon
Winnipeg
Ignace
Kapuskasing
Kirkland Lake
Jonquière
Sept-Îles
ST LAWRENCE
Gander
Bellingham
Chilliwack
Thunder Bay
Longlac
Chicoutimi
Victoria
Coeur d'Alene
Havre
Minot
Sudbury
Rouyn
Québec
Corner Brook
Seattle
Spokane
Williston
Grand Forks
North Bay
Trois-Rivières
Riverview
St John's
Olympia
Yakima
Helena
Great Falls
Bismarck
Fargo
Duluth
Sault Sainte Marie
Montréal
Charlottetown
COLUMBIA
MISSOURI
Marquette
Scarborough
Fredericton
Butte
Billings
Dickinson
St Cloud
Ottawa
Portland
Bozeman
Aberdeen
Minneapolis
St Paul
Hull
Montpelier
Bangor
Dartmouth
Halifax
Salem
Boise
Rapid City
Pierre
Eau Claire
Lansing
Toronto
Rochester
Augusta
Bend
Madison
Eugene
Twin Falls
Casper
Sioux Falls
Milwaukee
Detroit
Buffalo
Syracuse
Concord
Scottsbluff
Cedar Rapids
Erie
Albany
Boston
Redding
Ogden
Cheyenne
Chicago
Toledo
Cleveland
Scranton
Providence
Salt Lake City
North Platte
Des Moines
Fort Wayne
Pittsburgh
Sacramento
Sparks
Denver
Omaha
Columbus
New York
Philadelphia
Oakland
Reno
Provo
Lincoln
Aurora
Indianapolis
Cincinnati
OHIO
San Francisco
Carson City
COLORADO
Colorado Springs
Kansas City
Dover
Baltimore
San Jose
Pueblo
Dodge City
Topeka
Louisville
Washington D.C.
Fresno
Santa Fe
Wichita
Tulsa
St Louis
Richmond
Las Vegas
RIO GRANDE
Albuquerque
Oklahoma City
Springfield
Nashville
Norfolk
Bakersfield
Phoenix
Amarillo
Bentonville
MISSISSIPPI
Memphis
Knoxville
Raleigh
Los Angeles
Lubbock
Atlanta
Charlotte
San Diego
Glendale
Dallas
Little Rock
Birmingham
Charleston
Tijuana
Tucson
El Paso
Abilene
Fort Worth
Shreveport
Jackson
Savannah
Ensenada
Mexicali
Nogales
Ciudad Juárez
Austin
Waco
Montgomery
Baton Rouge
Mobile
Jacksonville
Rosario
Hermosillo
Múzquiz
San Antonio
Houston
New Orleans
Tallahassee
Chihuahua
Corpus Christi
Orlando
Bahía Tortugas
Monclova
Tampa
Los Mochis
Nuevo Laredo
Villa Insurgentes
Monterrey
Miami
La Paz
Torreón
Durango
Matamoros
San Lucas
Tepic
Ciudad Victoria
Nassau
THE BAHAMAS
TURKS AND CAICOS ISLANDS
Mazatlán
León
Havana
ANGUILLA
San Luis Potosí
Santa Clara
ST KITTS AND NEVIS
CUBA
VIRGIN ISLANDS
Basseterre
Guadalajara
Mérida
Guantánamo
DOMINICAN REPUBLIC
PUERTO RICO
ANTIGUA AND BARBUDA
Saint John's
Puebla
Campeche
CAYMAN ISLANDS
HAITI
Toluca
Mexico City
Villahermosa
Port au Prince
Santo Domingo
Acapulco
Chetumal
JAMAICA
MONTSERRAT
DOMINICA
Veracruz
Teapa
BELIZE
Kingston
Oaxaca
Tuxtla Gutiérrez
Belmopan
GAUDELOUPE
Roseau
ST LUCIA
MARTINIQUE
Castries
GUATEMALA
HONDURAS
Bridgetown
BARBADOS
Guatemala City
Tegucigalpa
ST VINCENT AND THE GRENADINES
EL SALVADOR
Kingstown
San Salvador
NICARAGUA
GRENADA
Managua
ARUBA
TRINIDAD AND TOBAGO
Saint George's
LAKE NICARAGUA
CURAÇAO
Port of Spain
COSTA RICA
Barranquilla
San José
Maracaibo
Caracas
Maturín
PANAMA
Cartagena
Valencia
Maracay
Panama City
Montería
VENEZUELA
Ciudad Bolívar
Medellín
San Cristóbal
Georgetown
New Amsterdam
Quibdó
Bucaramanga
Ibagué
Tunja
Puerto Ayacucho
GUYANA
Pereira
Buenaventura
Cali
Bogotá
Villavicencio
Ituni
Paramaribo
Kourou
COLOMBIA
Boa Vista
SURINAME
Cayenne
Popayán
Neiva
San José del Guaviare
FRENCH GUIANA
Calçoene
Pasto
Florencia
Maloca
Esmeraldas
Mitú
Boiaçu
Oriximina
Macapá
Afuá
Manta
Ibarra
Portoviejo
Quito
Tres Esquinas
NEGRO
Marañ
Óbitos
Castanhal
ECUADOR
Belém
Braganca
Jipijapa
Cabo Pantoja
Fonte Boa
Tauapeçaçu
AMAZON
Riobamba
Urucurituba
Santarém
São Luís
Guayaquil
Iquitos
Tonantins
Altamira
Parnaíba
Cuenca
Jutaí
Manaus
Tucuruí
Sobral
Fortaleza
Piura
Moyobamba
Carauari
São Félix
Quixeramobim
Sullana
Tarapoto
Eirunepé
Xinguara
Teresina
Natal
Chiclayo
PURUS
MADEIRA
Jacareacanga
Maraba
Guarabira
Cruzeiro do Sul
Tapauá
BRAZIL
Araguaína
Imperatriz
Mombaça
João Pessoa
Trujillo
Pucallpa
Feijó
Uruçuí
Floresta
Chimbote
Porto Velho
Alta Floresta
Petrolina
Recife
PERU
Rio Branco
Ariquemes
Xique Xique
Juàzeiro
Jaboatão
Callao
Brasileia
Riberalta
Jarú
Palmas
Maceió
Lima
Camisea
Costa Marques
Ji-Paraná
Barreiras
Aracaju
Huancayo
Puerto Maldonado
Pimenta Bueno
Feira de Santana
Ayacucho
Santa Ana de Yacuma
Salvador
Cusco
Vilhena
Formosa
Jequié
Itabuna
Ica
Juliaca
LAKE TITICACA
Brasilia
Montes Claros
Ilhéus
La Paz
Trinidad
Cuiabá
Cáceres
Goiânia
Teófilo Otôni
Arequipa
BOLIVIA
Rio Verde
Patos de Minas
Aracruz
Cochabamba
Santa Cruz
Uberaba
Araxá
Ipatinga
Arica
Sucre
Campo Grande
Caçu
Belo Horizonte
Iquique
Potosí
Ponto Porã
Araçatuba
Vitória
Tocopilla
Tarija
Pedro Juan Caballero
Ribeirão Preto
Juiz de Fora
São João da Barra
Órán
Salta
PARAGUAY
Marília
Botucato
Nova Iguaçu
Antofagasta
San Salvador de Jujuy
Apucarana
Campinas
Rio de Janeiro
Maringá
San Miguel de Tucumán
Asunción
Itararé
São Paulo
Curitiba
Santiago del Estero
Formosa
Foz do Iguaçu
Copiapó
Resistencia
Encarnación
Corrientes
Florianópolis
Coquimbo
Catamarca
Posadas
La Rioja
Reconquista
Santa Maria
Santa Fé
Porto Alegre
San Juan
Paraná
Concordia
Rio Grande
Quilpué
Córdoba
Paysandú
Viña del Mar
Mendoza
Treinta y Tres
URUGUAY
Rosario
Rocha
Valparaíso
San Luis
Zárate
Montevideo
Maldonado
Santiago
PARANÁ
SÃO FRANCISCO
ARAGUAIA
Aberdeen
Dundee
Glasgow
Edinburgh
Leeds
Belfast
Manchester
IRELAND
Liverpool
Dublin
UNITED KINGDOM
Kilkenny
Birmingham
Cork
Cardiff
London
Antwerp
Ghent
Brussels
BELGIUM
Amiens
Rouen
Caen
Paris
Brest
Orléans
Angers
FRANCE
Nantes
Bourges
Limoges
St-Étienne
Bordeaux
Lyon
Marseille
A Coruña
Bilbao
Toulouse
Vigo
Pamplona
ANDORRA
Andorra la Vella
Ourense
Zaragoza
Barcelona
SPAIN
Oporto
Valencia
Madrid
Palma de Mallorca
Lisbon
Toledo
Murcia
BALEARIC ISLANDS
PORTUGAL
Cartagena
Seville
Córdoba
Granada
Algiers
Oran
Tangier
Oujda
Ech Chèlif
Rabat
Sidi Bel Abbès
Fès
Casablanca
Figuig
Aïn Sefra
MOROCCO
Béchar
Beni Mellal
Abadla
Marrakesh
Ouarzazate
Agadir
El Goléa
Tiznit
Bou Izakarn
Timimoun
Tarfaya
ALGERIA
Laâyoune
Tindouf
Adrar
Es Samara
In Salah
WESTERN SAHARA
Chegga
Reggane
Bir Mogreïn
Ad Dakhla
Zouérat
Nouâdibou
Atâr
Araouane
Akjoujt
Boughessa
MAURITANIA
Tidjikja
Ayoûn el Atroûs
Nouakchott
MALI
Anéfis
St-Louis
Kiffa
Timbuktu
SÉNÉGAL
Goundam
Gao
Dakar
Nioro
NIGER
SENEGAL
Mopti
Dori
Kaolack
Kayes
Ségou
Gourcy
Niamey
Banjul
THE GAMBIA
Koudougou
Bissau
Labé
Bamako
Ouagadougou
GUINEA-BISSAU
Bobo-Dioulasso
Kindia
GUINEA
Kankan
BURKINA FASO
Conakry
Korhogo
Wa
Tamale
BENIN
Port Loko
Sokodé
Parakou
Freetown
CÔTE D'IVOIRE
GHANA
Iseyin
SIERRA LEONE
Bouaké
Kumasi
TOGO
Monrovia
Yamoussoukro
Lomé
Porto-Novo
LIBERIA
Abidjan
Accra
Cape Coast

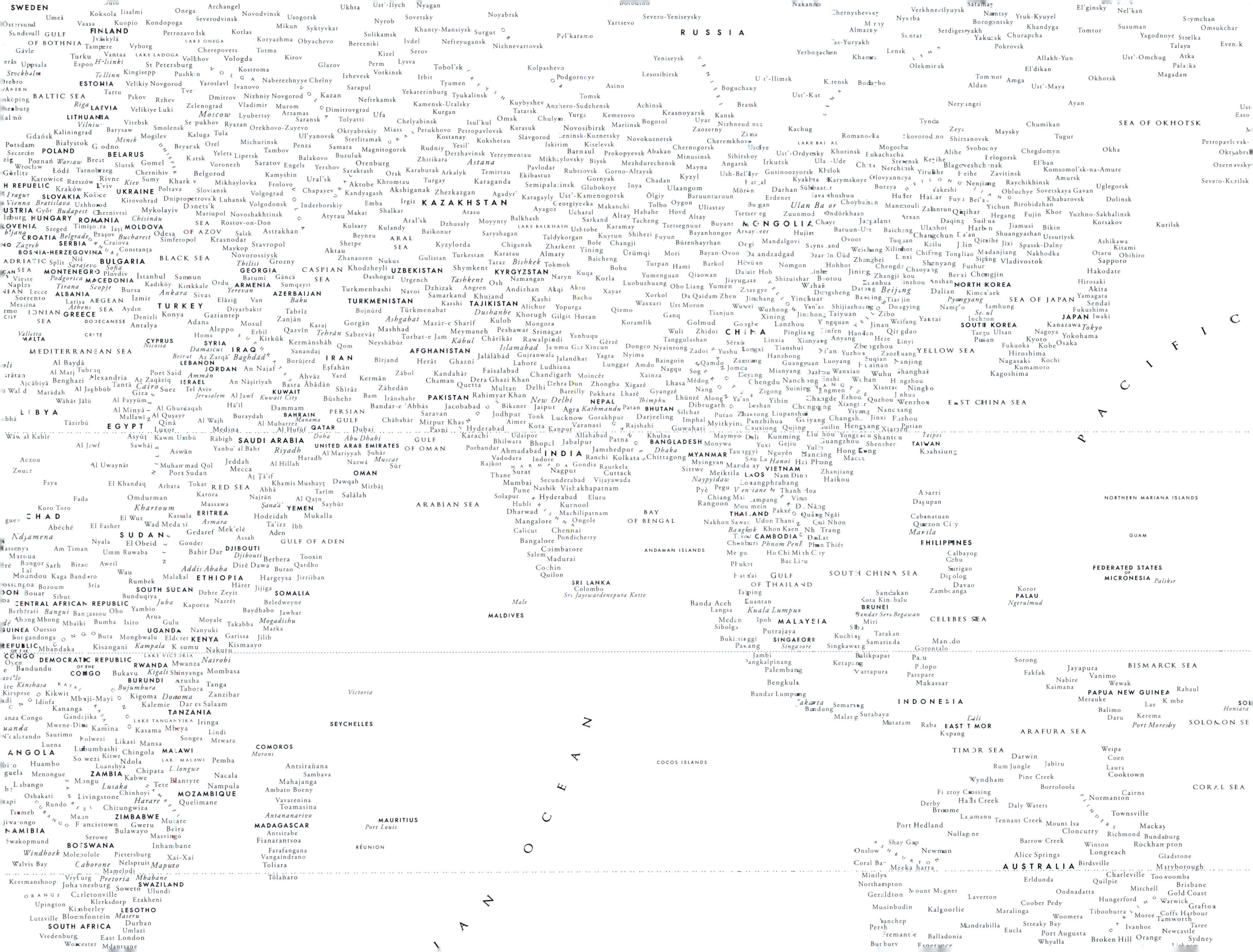

SWEDEN
FINLAND
GULF OF BOTHNIA
RUSSIA
ESTONIA
BALTIC SEA
LATVIA
LITHUANIA
POLAND
BELARUS
UKRAINE
SLOVAKIA
HUNGARY
ROMANIA
MOLDOVA
CROATIA
SERBIA
BOSNIA-HERZEGOVINA
BULGARIA
MONTENEGRO
MACEDONIA
ALBANIA
GREECE
TURKEY
BLACK SEA
SEA OF AZOV
GEORGIA
ARMENIA
AZERBAIJAN
CASPIAN SEA
KAZAKHSTAN
UZBEKISTAN
TURKMENISTAN
KYRGYZSTAN
TAJIKISTAN
ARAL SEA
MONGOLIA
CHINA
NORTH KOREA
SOUTH KOREA
JAPAN
SEA OF JAPAN
SEA OF OKHOTSK
YELLOW SEA
EAST CHINA SEA
TAIWAN
CYPRUS
SYRIA
LEBANON
ISRAEL
JORDAN
IRAQ
IRAN
KUWAIT
BAHRAIN
QATAR
UNITED ARAB EMIRATES
OMAN
SAUDI ARABIA
YEMEN
PERSIAN GULF
GULF OF OMAN
AFGHANISTAN
PAKISTAN
INDIA
NEPAL
BHUTAN
BANGLADESH
MYANMAR
LAOS
VIETNAM
THAILAND
CAMBODIA
PHILIPPINES
MEDITERRANEAN SEA
LIBYA
EGYPT
RED SEA
ARABIAN SEA
BAY OF BENGAL
GULF OF ADEN
CHAD
SUDAN
SOUTH SUDAN
ERITREA
DJIBOUTI
ETHIOPIA
SOMALIA
CENTRAL AFRICAN REPUBLIC
DEMOCRATIC REPUBLIC OF THE CONGO
UGANDA
KENYA
RWANDA
BURUNDI
TANZANIA
ANGOLA
ZAMBIA
MALAWI
MOZAMBIQUE
ZIMBABWE
NAMIBIA
BOTSWANA
SOUTH AFRICA
SWAZILAND
LESOTHO
MADAGASCAR
COMOROS
SEYCHELLES
MAURITIUS
RÉUNION
MALDIVES
SRI LANKA
ANDAMAN ISLANDS
COCOS ISLANDS
SOUTH CHINA SEA
GULF OF THAILAND
MALAYSIA
BRUNEI
SINGAPORE
CELEBES SEA
INDONESIA
EAST TIMOR
TIMOR SEA
ARAFURA SEA
PALAU
FEDERATED STATES OF MICRONESIA
NORTHERN MARIANA ISLANDS
GUAM
PAPUA NEW GUINEA
BISMARCK SEA
SOLOMON SE
CORAL SEA
AUSTRALIA
PACIFIC
INDIAN OCEAN

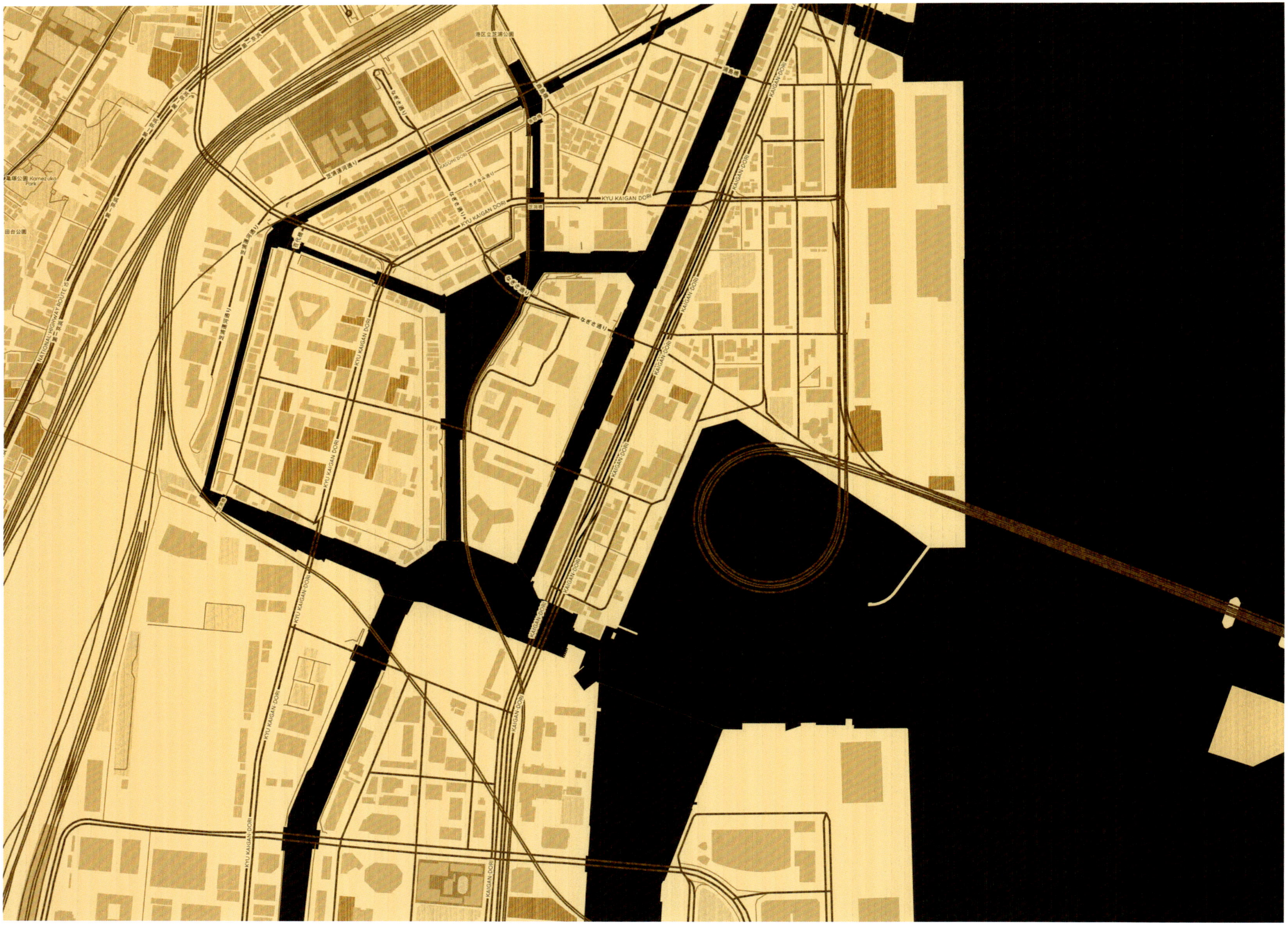

KYU KAIGAN DORI
KAIGAN DORI
KAISHI DORI
なぎさ通り
芝浦運河通り
NATIONAL HIGHWAY ROUTE 15
Kamezuka Park
港区立芝浦公園

« Digital Woodcut

Eleanor ***Lutz*** *— Mapbox*

I made this map during a week-long visit to the offices of Mapbox in Washington, DC. Mapbox runs an open-source mapping platform, and their team was teaching me how to use their mapmaking software. I wanted to try something really fun as a learning experiment, so I decided to recreate a traditional woodwork map using only digital tools. My goal was to create a "handmade" style template that could be applied to every part of the world as a customized map skin.

To do this, I used the Mapbox program TileMill 2 to add different wooden textures to everything from parking lots and hospitals to wetlands and forests. The original textures were made by PixEden, and I color-corrected, resized, and smoothed the files to fit seamlessly into every corner of the map. I also added a rich, red-orange tone to all of the wooden tiles to tie everything together. The roads and rivers are also made with tiny, thin lines patterned to look like wood.

The most technically difficult part of the map was the wooden placard shapes I added to the background of each country name. I had to make sure that each placard fit the length of the country name, and also make sure that none of the placards overlapped one another at any zoom level. To do this I made a set of several preset placard sizes, and used a small script written in Java to assign each country to the right size placard by name length.

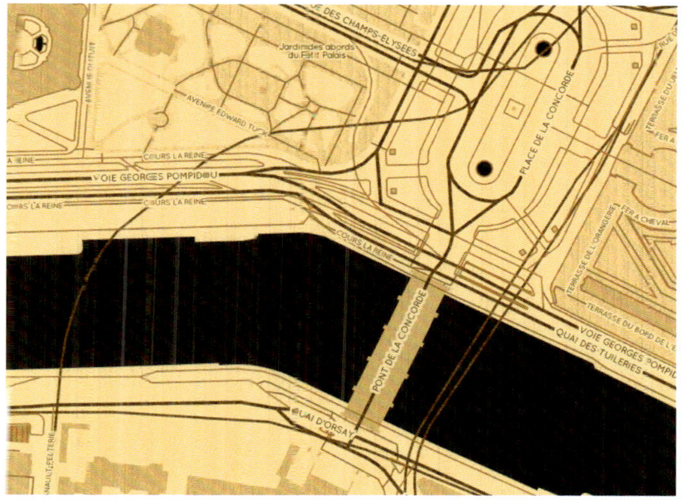

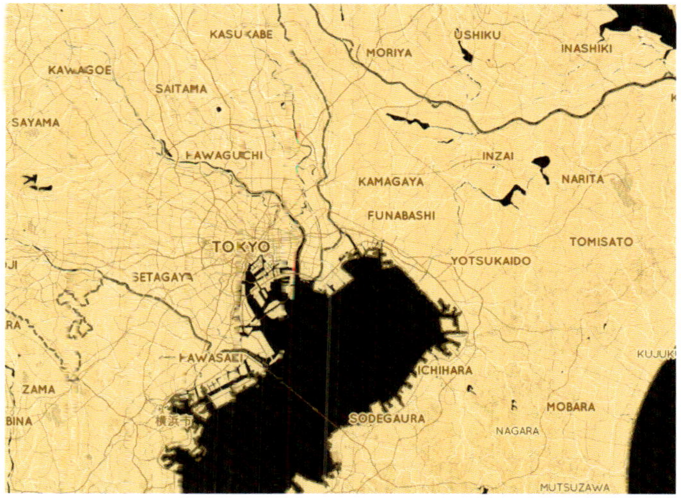

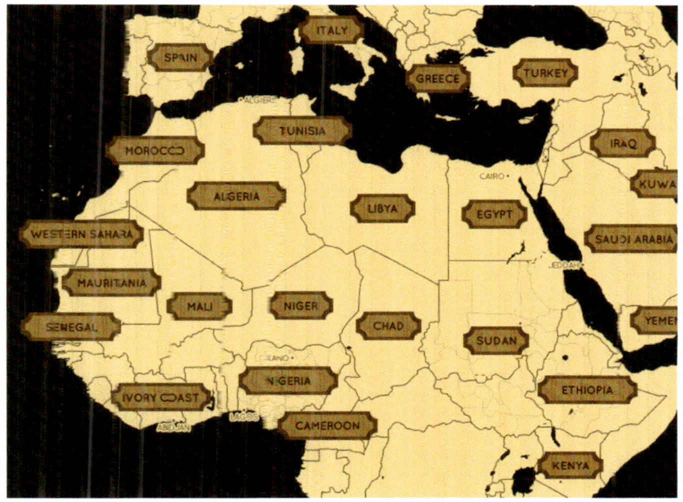

» Rather by accident: that's how I rolled into cartography.

New York »

KHUAN+KTRON

Rather by accident: that's how I rolled into cartography. I'd been working as a commercial illustrator for an odd ten years when I got a gig drawing weekly cover illustrations for the summer issues of a Belgian magazine. Every new issue would focus on another European country, and I was asked to design a series of covers that would make these countries instantly recognizable. After a first round of sketches, the staff settled for my simplest idea: I was to draw a map of each country showing its most well-known cities, as well as a fine selection of the nation's most amusing clichés. The series turned out pretty good: the client was happy, and response was overwhelmingly positive. Soon after it was published, commissions started pouring in, and I found myself increasingly drawing more maps, all the while keeping a strong emphasis on the illustrative aspect of cartography. Since then, I've drawn countries and regions from all over the globe, not to mention the occasional world map here and there.

This particular map was commissioned by *The Good Life* magazine to brighten up the contents page of their New York-themed issue. The brief was, well... pretty brief: "draw us a map of New York with the airports on it, and this list of buildings and locations we will be featuring in the magazine." I'd never been to the Big Apple, so I parked myself behind my computer with my little list and started doing research, amassing virtual piles of reference material. Once I got the city's outlines straight and the main streets were in the right place (thank you, Google Maps), it was pretty straightforward a task: adding detail, then another layer of detail, and on top of that... more detail.

When you're doing an editorial illustration, it doesn't matter too much what you are drawing exactly—be it a map or a bunch of figures/things/situations, or whatever the brief says. But it does have to *fit*: there's no room for falseness and no margin for errors. One of the lessons that has always stuck with me is this one: "whatever it is you're drawing, when a specialist looks at your picture he/she has to be happy and content." And when it comes to cartography, *everybody* is a specialist.

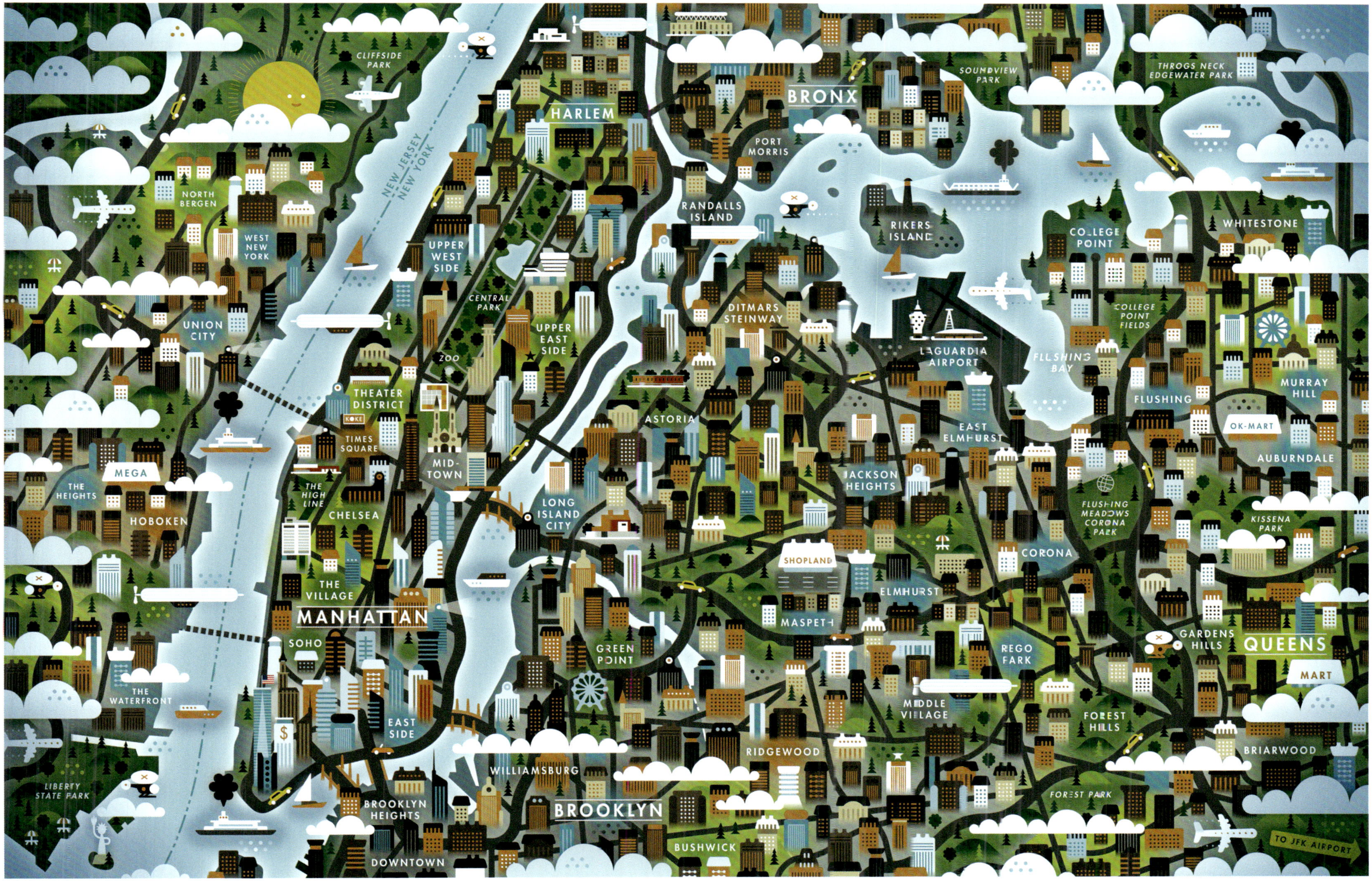
CLIFFSIDE PARK
HARLEM
BRONX
SOUNDVIEW PARK
THROGS NECK EDGEWATER PARK
NEW JERSEY
NEW YORK
NORTH BERGEN
WEST NEW YORK
PORT MORRIS
RANDALLS ISLAND
RIKERS ISLAND
COLLEGE POINT
WHITESTONE
UPPER WEST SIDE
CENTRAL PARK
UPPER EAST SIDE
DITMARS STEINWAY
LAGUARDIA AIRPORT
COLLEGE POINT FIELDS
FLUSHING BAY
UNION CITY
ZOO
THEATER DISTRICT
ASTORIA
FLUSHING
MURRAY HILL
OK-MART
TIMES SQUARE
EAST ELMHURST
AUBURNDALE
MEGA
MID-TOWN
THE HEIGHTS
THE HIGH LINE
JACKSON HEIGHTS
LONG ISLAND CITY
FLUSHING MEADOWS CORONA PARK
HOBOKEN
CHELSEA
KISSENA PARK
CORONA
SHOPLAND
THE VILLAGE
ELMHURST
MANHATTAN
MASPETH
SOHO
GREEN POINT
REGO PARK
GARDENS HILLS
QUEENS
MART
THE WATERFRONT
MIDDLE VILLAGE
EAST SIDE
FOREST HILLS
RIDGEWOOD
BRIARWOOD
WILLIAMSBURG
LIBERTY STATE PARK
BROOKLYN HEIGHTS
BROOKLYN
FOREST PARK
BUSHWICK
DOWNTOWN
TO JFK AIRPORT

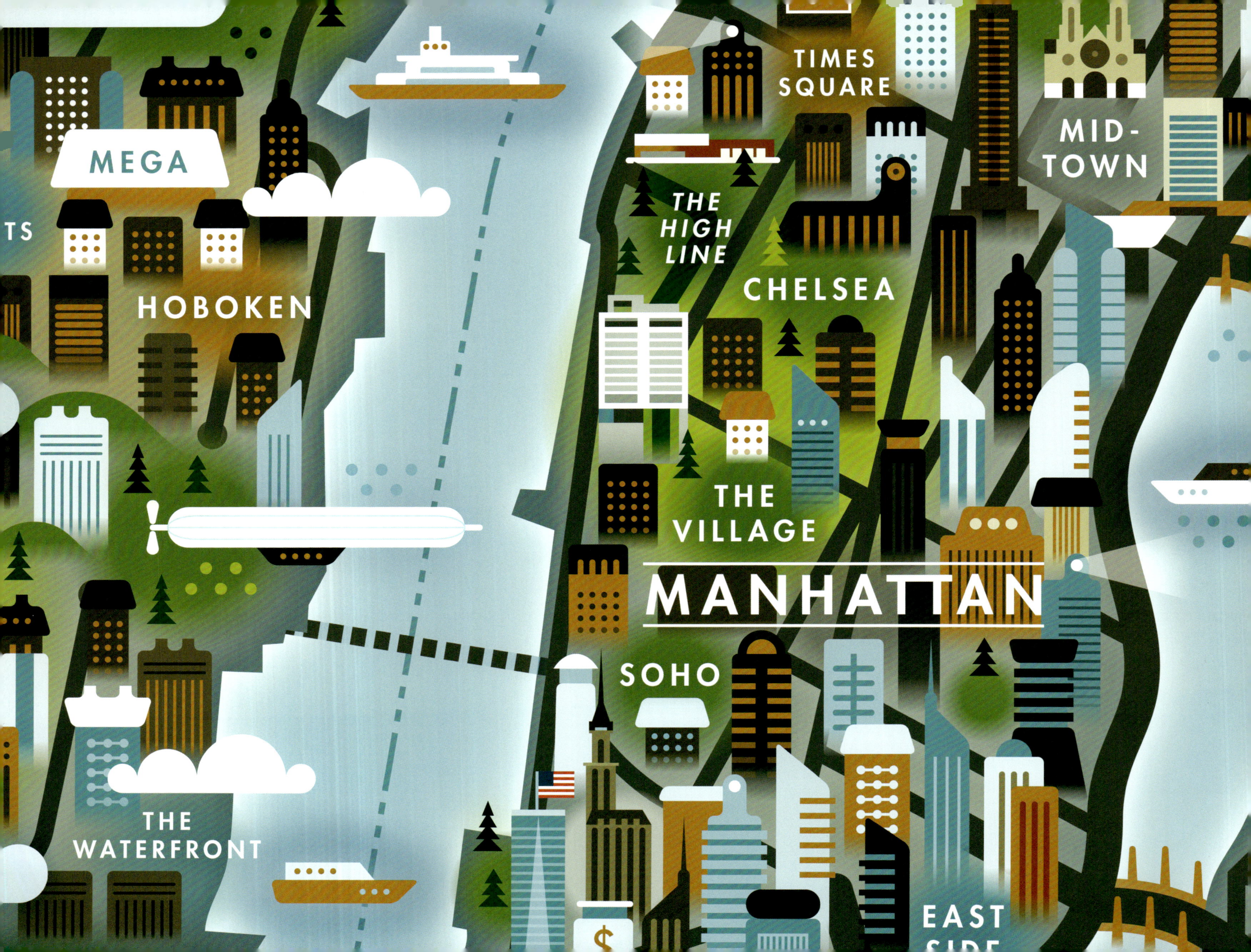
MEGA
TS
HOBOKEN
THE WATERFRONT
TIMES SQUARE
MID-TOWN
THE HIGH LINE
CHELSEA
THE VILLAGE
MANHATTAN
SOHO
EAST

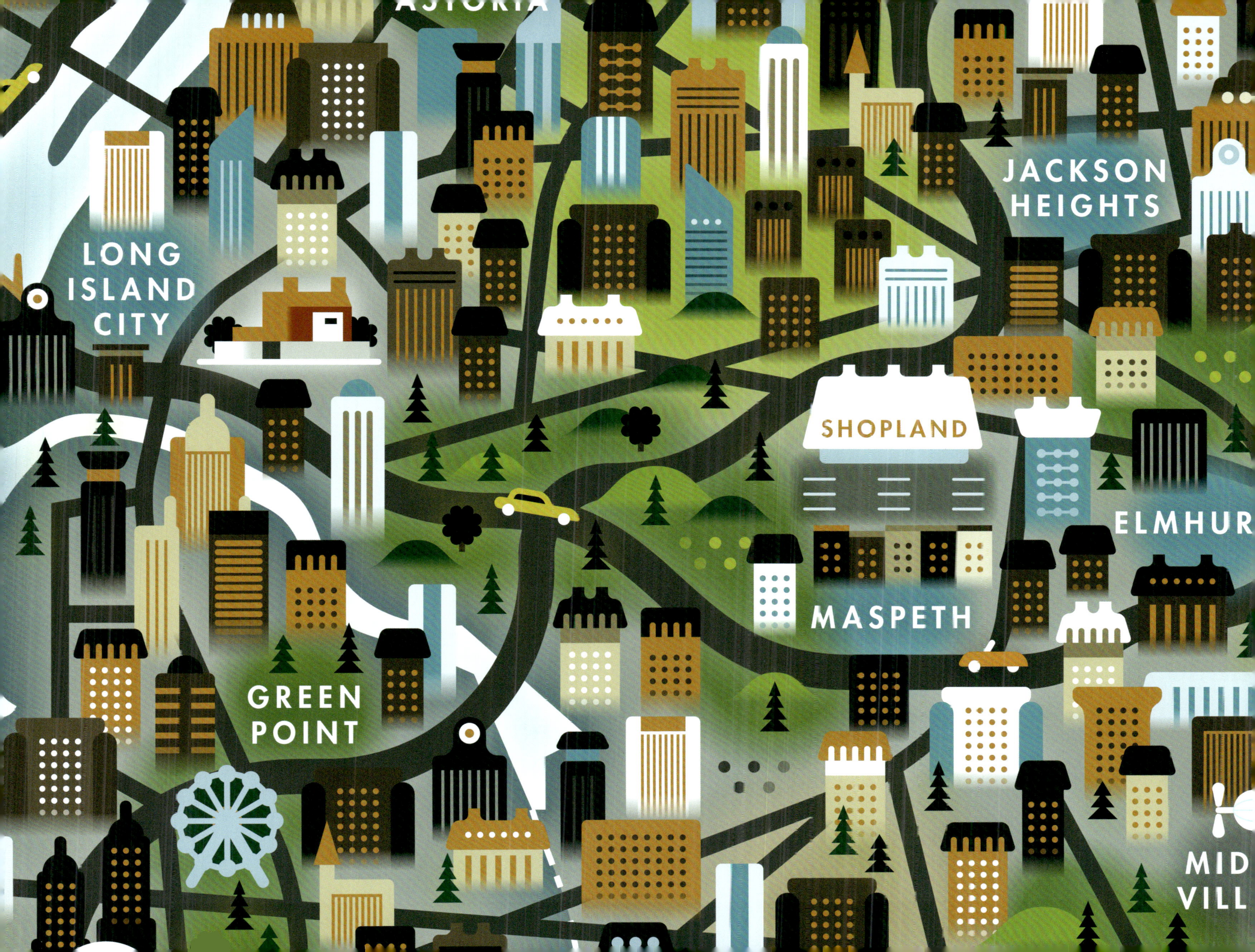
ASTORIA
LONG ISLAND CITY
JACKSON HEIGHTS
SHOPLAND
ELMHUR
MASPETH
GREEN POINT
MID
VILL

USW findaroom

Find

g219 is a lecture theatre located on the second floor of the Glynneath building on the Treforest campus.

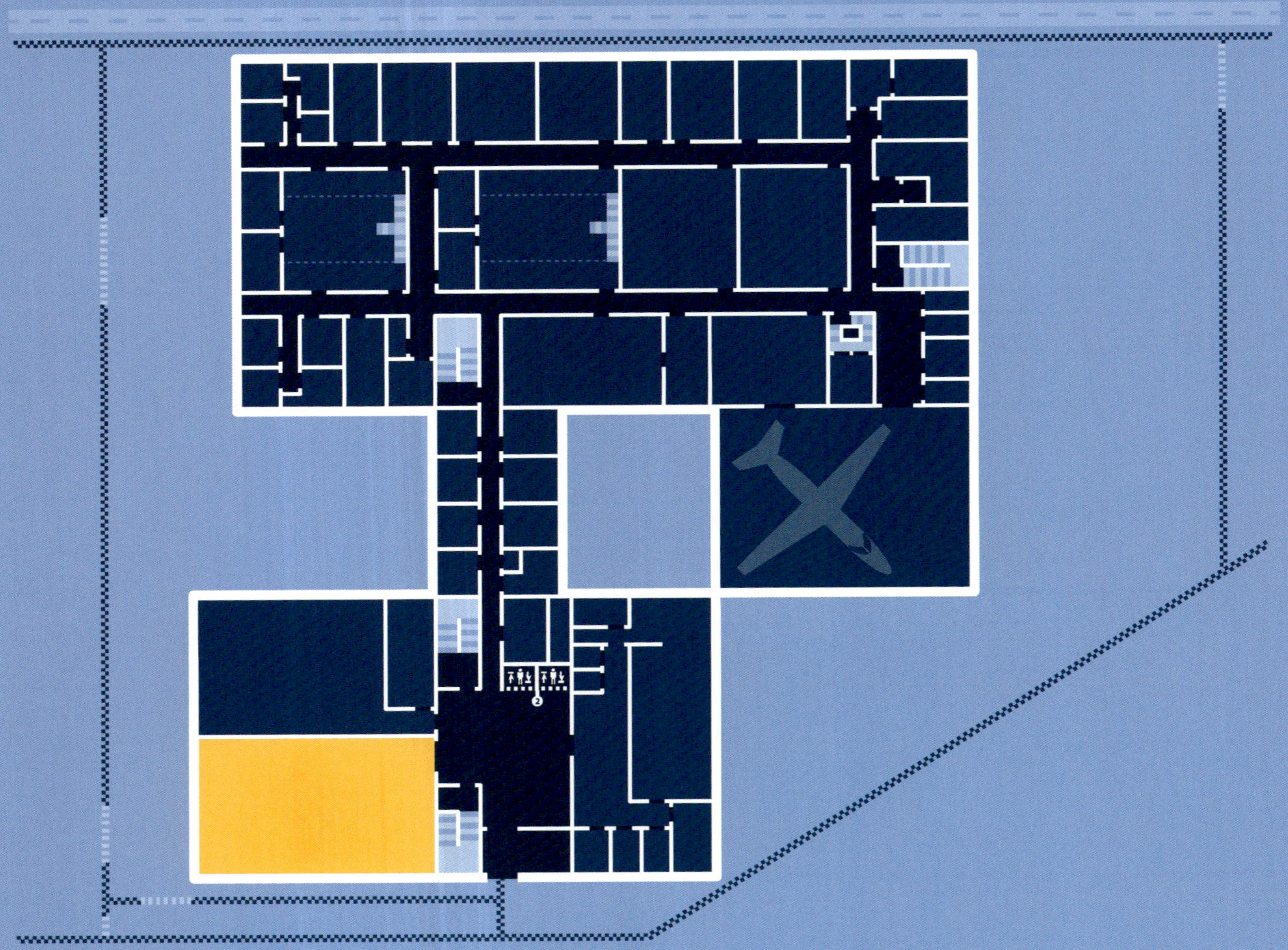

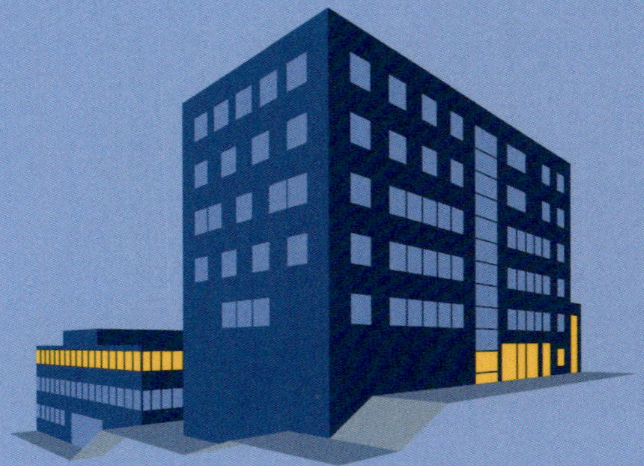

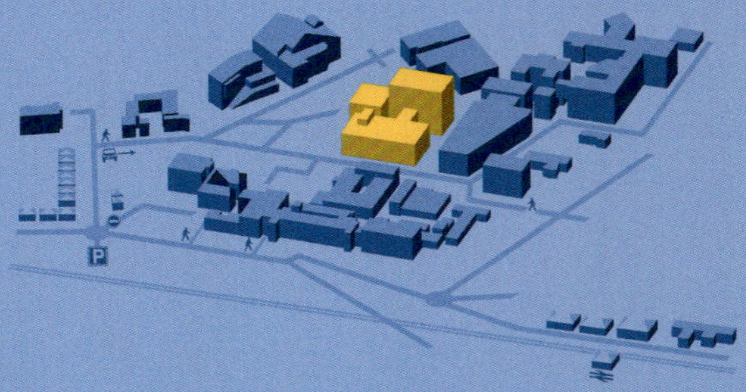

Map Legend

FindARoom is a service from the University of South Wales

《 FindARoom—University of South Wales

Barry ***Richards***

Locating a room at the University of South Wales (USW) is inherently challenging. USW has eight campuses on five different sites across the region, with many large buildings on the sides of steep hills. Staff and students alike find it difficult to locate rooms. *FindARoom* delivers an elegant, visual, browser-based solution to this problem.

From the start I set out to convey information on the maps using as few colours as possible. I chose blue as the dominant colour as it is a flexible colour that can be made light or dark while remaining recognisably blue; in contrast red, for example, becomes either pink or brown. I restricted myself to seven shades of blue across the four different types of maps used, which created a powerful style that unifies the maps. When used with a strong yellow for highlighting rooms, floors, and buildings, a high level of contrast is achieved, which enables users (particularly the visually impaired) to quickly locate their destination on the maps. The palette is also a reference to an architect's blueprint.

The use of three different views of the same destination is crucial in helping users understand its location. The campus-level map shows the position of the building, and in conjunction with the building projection, aids navigation by enabling landmark-based wayfinding. The projection also helps users to understand buildings where the ground floor may not obviously be on the ground. Both of these are subordinate to the floor plan, which locates the room within the building, and only includes information necessary for wayfinding and creating a mental model of the building's layout.

As Wales is a bilingual country, and USW has many international students, I made the decision to exclude text in favour of icons or visual language wherever possible on the maps.

The maps are displayed to users using SVG (Scalable Vector Graphics). This means that they scale to any screen size without loss of image quality and look especially good on high-resolution Retina displays.

From the start I set out to convey information on the maps using as few colours as possible.

findaroom

Find

gt8017 is a teaching room located on the ground floor of the Alfred Russel Wallace building on the Upper Glŷntaff campus.

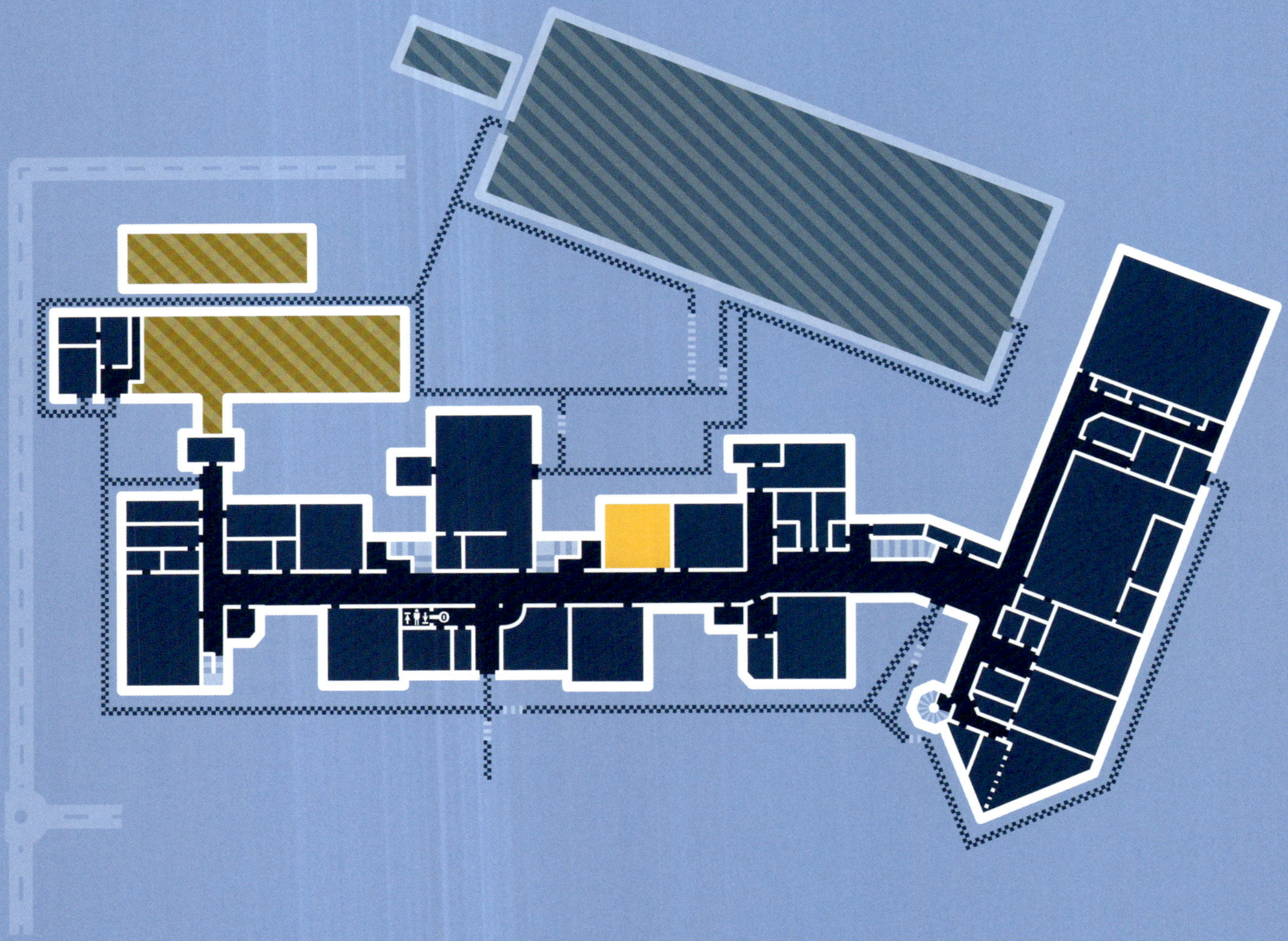

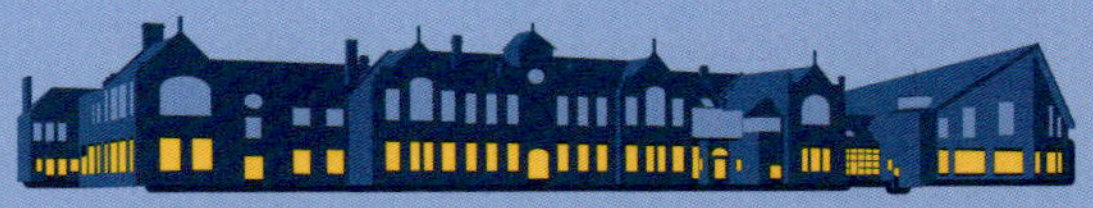

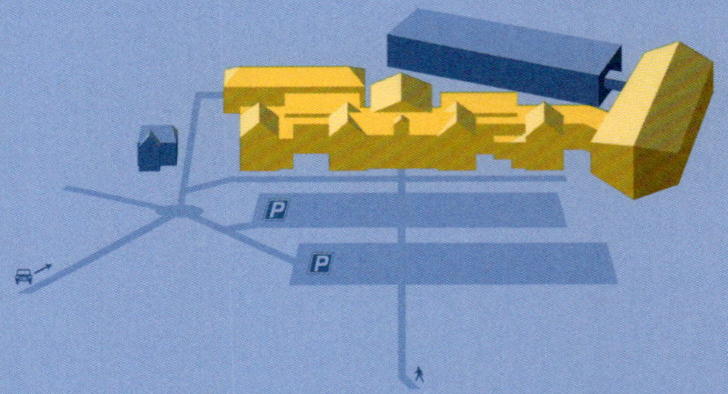

Map Legend

FindARoom is a service from the University of South Wales

USW

findaroom

Find

ccam01 is a teaching room located on floor AM of City Campus, Newport.

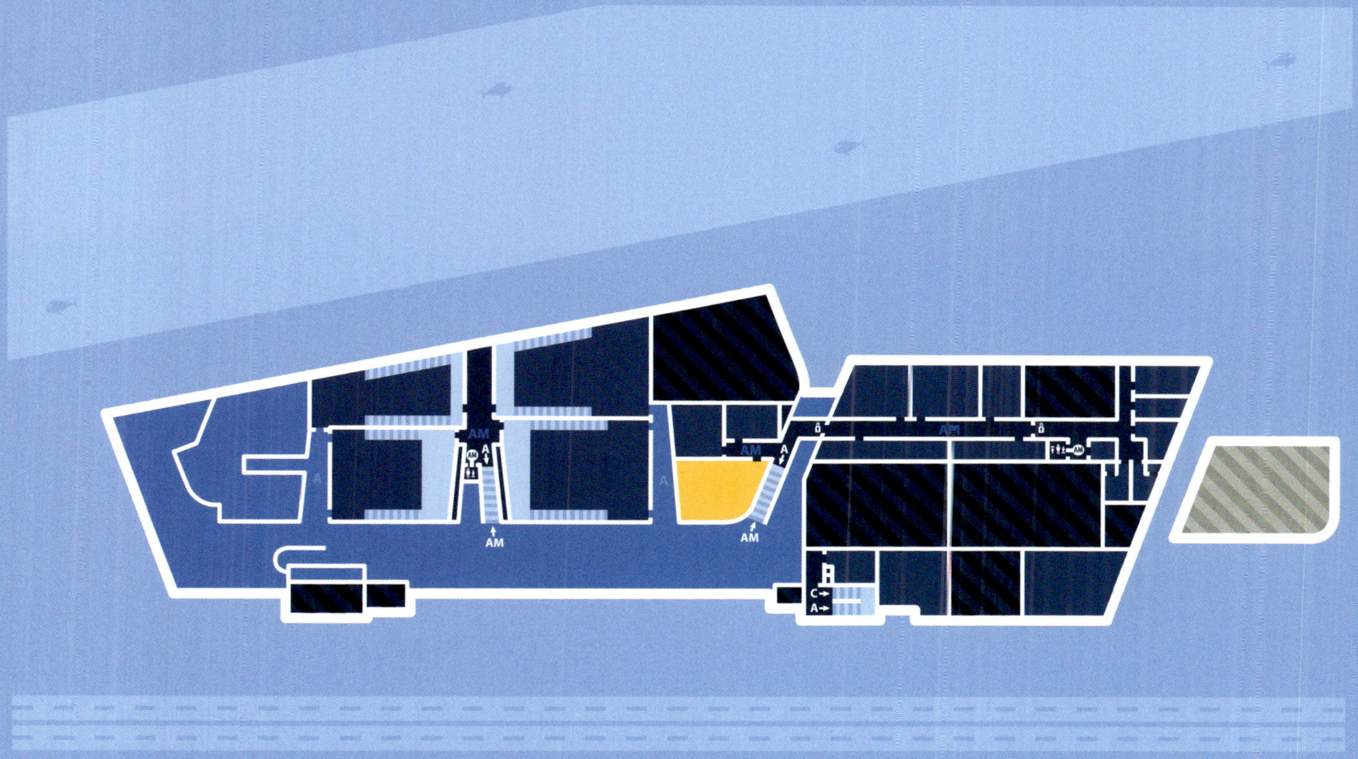

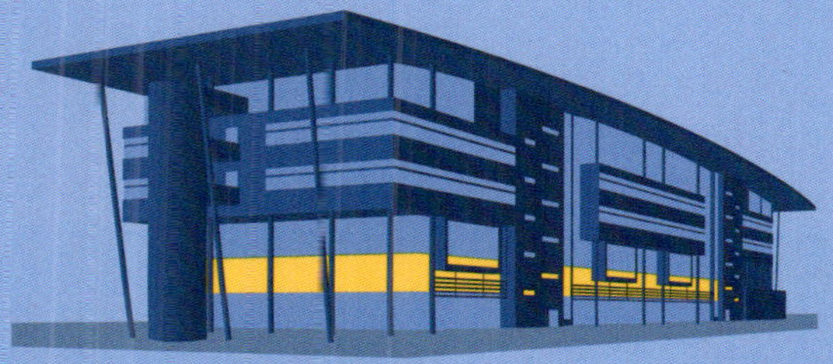

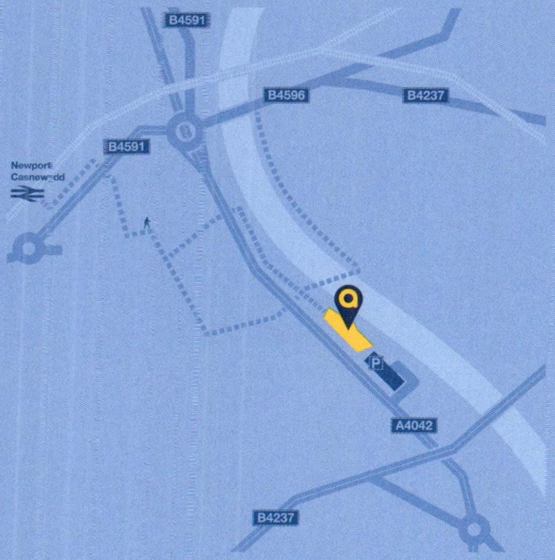

Map Legend

FindARoom is a service from the University of South Wales

atlasofdesign.org